# NEXT GENERATION OPTICAL NETWORKS

The Convergence of IP Intelligence
and Optical Technology

ISBN 0-13-028226-X

90000

9 780130 282262

## Prentice Hall Series in
## Computer Networking and Distributed Systems

*Radia Perlman, Series Editor*

# NEXT GENERATION OPTICAL NETWORKS

The Convergence of IP Intelligence
and Optical Technology

Peter Tomsu • Christian Schmutzer

Prentice Hall PTR
Upper Saddle River, NJ 07458
www.phptr.com

Library of Congress Catalog Card Number: 2001033944

Editorial/production supervision: *Donna Cullen-Dolce*
Acqusition Editor: *Mary Franz*
Editorial Assistant: *Noreen Regina*
Marketing Manager: *Dan DePasquale*
Manufacturing Manager: *Alexis R. Heydt*
Cover Design Director: *Jerry Votta*
Interior Design: *Gail Cocker-Bogusz*

 © 2002 Prentice Hall PTR
Prentice-Hall, Inc.
Upper Saddle River, NJ 07458

The publisher offers discounts on this book when ordered in bulk quantities.
For more information, contact
Corporate Sales Department,
Prentice Hall PTR
One Lake Street
Upper Saddle River, NJ 07458
Phone: 800-382-3419; FAX: 201-236-714
E-mail (Internet): corpsales@prenhall.com

Printed in the United States of America

10 9 8 7 6 5 4 3 2 1

ISBN 0-13-028226-X

Prentice-Hall International (UK) Limited, *London*
Prentice-Hall of Australia Pty. Limited, *Sydney*
Prentice-Hall Canada Inc., *Toronto*
Prentice-Hall Hispanoamericana, S.A., *Mexico*
Prentice-Hall of India Private Limited, *New Delhi*
Prentice-Hall of Japan, Inc., *Tokyo*
Prentice-Hall (Singapore) Pte. Ltd., *Singapore*
Editora Prentice-Hall do Brasil, Ltda., *Rio de Janeiro*

*To our family and friends*

# Contents

## Chapter 2

## Chapter 3

# Optical Networking Technology Fundamentals  67

## Chapter 4

# Existing and Future Optical Control Planes    187

## Chapter 5

# Optical Networking Applications and Case Examples   267

# Acknowledgments

There are a lot of people we would like to thank. This book would not have been possible without the patience and support of our families. They provided sustained encouragement during the complete period of writing, from the very beginnings of planning this book through the final edits.

A special thanks goes to Mary Franz, who was a significant help in all phases of the project. Radia Perlman deserves a special recognition. She provided encouragement, ideas, guidance, and support throughout the entire effort, giving valuable input in structuring the book.

A very special thanks to our employer, Cisco Systems, for providing great help over the past two years, especially Jane Butler and Rob Lloyd, who supported this book from the very beginning. We also received great support from Gerhard Wieser, David Tsiang, George Suwala, and all others who helped us to finalize this book and are not listed here.

Many of the figures were adapted from Cisco Systems presentations, as well as ITU, IETF, ATM Forum, and other standardization bodies documents.

# Preface

## Why This Book

We had the idea to write this book two years ago, when service providers started commonly thinking about optical technologies such as Wavelength Division Multiplexing (WDM) to increase the capacity of optical transmission systems. The need for higher bandwidth came, of course, from the growing demand generated by the Internet and all of its applications and services.

Historically, there has not been much consensus between service providers and telecommunications camps (on the one side) and Internet service providers (ISPs) and Internet users (on the other side). Obviously, the most important reason for that is that, in the past, telephony companies built Time Division Multiplexing (TDM) networks and delivered voice services to their customers. As the Internet became more popular and generated increasing amounts of traffic, new ISPs started building router networks running IP. These delivered e-mail and World Wide Web (WWW) access and, later, more advanced IP services to their customers. Also at this time, traditional telephone providers started building router networks delivering IP services like their counterparts, the ISPs. From an organizational point of view, this represents two completely separated groups of people, two completely separated business functions, two completely separated areas of responsibility, two completely separated budgets, and—last but not least—two different human generations.

However, time changed quickly and while market deregulation was introduced, a broad range of alternate service providers started to compete in a very fast-growing market with the incumbent providers. This very competitive situation now forces every single service provider to optimize its network architecture and to streamline its internal organizational structure in order to minimize its operational expenses (OPEX), one of the most important keys to success.

Suddenly, people talking about IP are also talking about WDM and the other way around. Although this is fine and a major step forward, there is still the problem that most people who have worked in the transmission space for years are experts on TDM, WDM, and all other optical transmission technologies. On the other hand, people who have worked for years in the Internet arena are experts in data networking, IP, and IP routing protocols. Both are experts but only in their tightly focused area of technology.

This is where our book is intended to enter the arena—to build a bridge between the two historically separated technology areas. It first outlines the most important technologies of optical transmission space, then focuses on important IP networking technologies. Finally, it concentrates on the convergence of both optical and IP to conclude. We are sure that the reader will obtain a deep understanding of what is really behind the "IP + optical" topic, a topic already of significant interest today and one that will gain even more momentum during the next few years.

## Targeted Audience

The target audience for this book includes all people interested and working in the networking arena, service providers, and enterprises that are planning, designing, or deploying advanced optical IP-based infrastructures. The book is structured and written in a way that should be easily understandable for newcomers. It supplies many helpful references to standards and other literature and in-depth explanations, including information for advanced readers who want to get a complete picture of the state of the art of optical networking and the according control and provisioning mechanisms for these networks.

## Structure of This Book

Chapter 1 of this book gives an overview about existing and possible future carrier network architectures. The architectural elements of traditional multilayer networks are described, and essential information about applied technologies such as Synchronous Optical Network/Synchronous Digital Hierarchy (SONET/SDH), Asynchronous Transfer Mode (ATM), or Multiprotocol Label Switching (MPLS), is included.

Chapter 2 focuses on standardization activities in the optical networking arena. A snapshot of what already has been specified by such standardization bodies as International Telecommunications Union (ITU), Institute of Electrical and Electronics Engineering (IEEE), Optical Internetworking Forum (OIF), or Internet Engineering Task Force (IETF) and what kind of proposals are to be finalized soon is provided.

Chapter 3 represents a technology backgrounder delivering in-depth knowledge about optical transmission technologies, data transmission technologies, and network survivability concepts required for understanding how next-generation networks are designed and deployed.

Chapter 4 concentrates on the overall network architecture of advanced optical IP-based infrastructures and describes the three major steps in the evolution, from static IP over optical networks to dynamic and integrated IP + optical networks.

Chapter 5 finally outlines optical end-to-end networking design trends seen in the industry. Three case examples are described to put the theory about the convergence of IP and optical technologies covered in Chapters 1 and 4 into a practical perspective.

## Further Information Online

All readers are welcome to consult the Web site www.nextgenzone.net for ongoing updated information related to this book and the authors, as well as for reading about the hot topics in the networking industry. A lot of up-to-date background information, including links to useful Web sites, industry articles, and much more online material can be found.

# 1

# Introduction to Carrier Network Architectures

In the first chapter, we give an overview about existing and possible future Carrier Network Architectures. We will explain the traditional architectural elements and describe the basic concepts and requirements of these architectures in order to put each of the different elements and its functionality into context. We do not go deep into the details of all used technologies, but rather point out the most important basics to provide the reader with just the necessary level of knowledge to understand the more detailed discussions of next-generation optical networks throughout the rest of the book.

## IP as Unifying Protocol Infrastructure

The explosive growth of Internet/intranet traffic is making its mark on the existing transport infrastructure. An unprecedented shift has occurred in traffic content, pattern, and behavior. It has transformed the design of multiservice networks and created a commercial demand for Internet Protocol (IP) networks that operate in excess of 1 Gigabit per second. A change in usage patterns from connection-oriented, fixed configured services to dynamic, connectionless IP services is currently underway. According to several studies, telephone company revenue will grow significantly, with data services—particularly IP—accounting for most of this increase. For public carriers, IP is critical for future revenue growth.

Exponential growth in IP traffic volumes is associated with continued strong growth of Internet and enterprise intranet network usage, rapid emergence of enhanced IP services based on Voice over IP (VoIP) and multicasting capabilities, and cost-effective, high-bandwidth residential connectivity via Digital Subscriber Line (DSL), cable modem, and wireless technologies.

From the networking perspective, IP is the only protocol that runs over any and all transport technologies. Even end-to-end solutions based on backbone technologies such as Asynchronous Transfer Mode (ATM) rely on external systems and applications to map IP traffic onto backbone circuits. Therefore, IP forms a common and standardized interface between services and the transport technologies that are used to deliver the services. Such a standardized interface enables operators to adapt quickly to rapidly changing markets, the introduction of new technologies, and increasing competition.

An intelligent network is one that is able to separate services from technology. Such an intelligent network will recognize individual users and applications, authenticate them as valid users of the network, enable them to select services, and deliver the appropriate level of performance to the applications.

Considering that the service providers' business of today is a mix of different services, such as dial access and Frame Relay, ATM has to be integrated. One way is to form an IP + ATM network infrastructure. An IP+ATM network infrastructure combines the application layer visibility of IP and the traffic management capabilities of ATM, all on a single platform. This enables service providers to provision services such as dial access and Frame Relay, as well as next-generation VoIP and integrated access services from a unified network architecture without compromising Quality of Service (QoS).

## The Traditional Carrier Network Architecture

Service providers have been using a mix of various kinds of networking technologies for building up their national or international carrier networks (see Figure 1–1). In doing so, they had to cope with several constraints and challenges. Each networking technology introduced by the service provider handled—and, in many networks, still handles—the issues of one or more of these challenges.

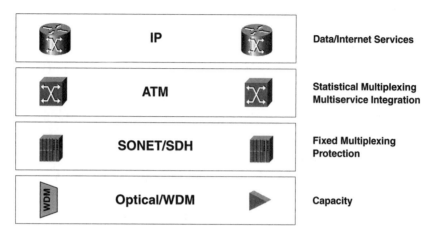

| | | |
|---|---|---|
| IP | | Data/Internet Services |
| ATM | | Statistical Multiplexing Multiservice Integration |
| SONET/SDH | | Fixed Multiplexing Protection |
| Optical/WDM | | Capacity |

**Figure 1–1**    The traditional network architecture consists of multiple layers

As a consequence, a traditional service provider network architecture is built of multiple layers. The optical/Wavelength Division Multiplexing (WDM) layer forms the physical transport medium providing sheer bandwidth; in the past, this layer did not have too much routing intelligence. To allocate bandwidth in a proper way, the Synchronous Optical Network (SONET)/Synchronous Digital Hierarchy (SDH) layer is used in many traditional networks. It offers mechanisms for efficient bandwidth utilization plus intelligent protection mechanisms but does not allow intelligent routing. The ATM layer above gives additional possibilities for statistical multiplexing while allowing multiservice integration at the same time. This basically enhances efficient utilization of the layers below (the SONET/SDH and the optical/WDM layer). ATM also defines routing mechanisms intended to optimize traffic delivery throughout the network in terms of the different ATM service offerings.

If service providers do not offer any IP services, the three-layer infrastructure described so far is more than sufficient. Because IP-based applications enforced IP service offerings, several mechanisms evolved integrating IP with the infrastructure described so far, ranging from pure overlay methods to very advanced integration models such as Multiprotocol Label Switching (MPLS).

This layered network architecture grew and changed as the services provided and sold by the providers changed and networking technology evolved. To understand this dramatic change in the architecture of today's networks fully, the functions and features of each of these layers will now be discussed in this chapter in more detail.

## Achieving Long-Haul Connectivity with Optical Transmission

Optical fiber as a transport medium offers a tremendous amount of capacity. We will look into optical transmission details in greater detail in the following chapters, but for the moment, we will try to keep it simple.

Because of the fibers' transmission capabilities, a modulated optical signal, which is usually inserted by a laser, can be transmitted several kilometers before it has to be recovered. With technologies available today (lasers as well as fiber), the signal has to be recovered every 40 to 80 km. The typical bit rates achievable with advanced modulation techniques today allow data to be sent at 10 Gbps per wavelength (optical frequency) and can be expected to go even higher as technology evolves. If multiple frequencies are used in parallel over one fiber (which generally is called **WDM**), the transmission capacity per fiber can be extended today into the Tb range.

By deploying these technologies, service providers can easily build optical networks with tremendous transmission capacity, ranging from several Gbps to Tbps, forming the basis for the current and next-generation backbone carrier networks.

## Delivering Multiplexed Services

In the past, service providers typically delivered telephony services to their customers. For this, they used nonsynchronous hierarchies called **Plesiochronous Digital Hierarchies** (PDHs) to carry the low-bit-rate signals representing either the customer's voice or data connections. PDH networks have several limitations. To begin with, the highest multiplexing rate is limited to 139.264 Mbps. In addition, the transport efficiency is low because there is a high amount of idle capacity required for accommodating delays between the nonsynchronous clocks of the network equipment. Furthermore, interoperability between different vendors is difficult because only multiplexing—not the transmission—is defined by the standards.

## Synchronous Transmission Standards

SONET and SDH overcome the drawbacks of PDH networks and enable service providers also to provision high-bit-rate connections above 155 Mbps. These connections are typically required for today's increasing data transport needs. SONET and SDH allowed service providers to build a Time Division Multiplexing (TDM) network on top of their physical fiber plant. Although SONET is the North American synchronous TDM standard, SDH is the TDM standard commonly used in Europe and Japan. Because SDH can be seen as a global standard (and SONET being a subset of SDH), interoperability at certain levels is ensured, as will be outlined in more detail later. In addition to the summary in this chapter, the Nortel whitepaper "SONET 101" [NORT-1], the Bellcore standard GR-253 [BELL-1], and the International Telecommunication Union (ITU) T recommendation G.707 [ITU-3] can be consulted for all the details on SONET/SDH.

Both SONET and SDH define a digital multiplexing hierarchy and should ensure compatibility of equipment and implement **synchronous networking**. The basic functionality is that client signals of different service types, such as E0, E1, DS0, T1, ATM, and others, are mapped into appropriate payloads that are then multiplexed into synchronous optical signals.

Both SONET and SDH accommodate nonsynchronous TDM hierarchies. SONET includes the North American hierarchy, which is based on the DS1 signal, combining 24 DS0s (56-Kbps channels) into one 1.54-Mb stream. SDH integrates the European hierarchy, which is based on the E1 signal, combining 32 E0s (64-Kbps channels) into one E1 at a speed of 2.048 Mbps, as can be seen from Table 1–1.

**Table 1–1**   The North American and European TDM Hierarchies

| NORTH AMERICAN RATE | | | EUROPEAN RATE | | |
|---|---|---|---|---|---|
| SIGNAL | SPEED | CHANNELS | SIGNAL | SPEED | CHANNELS |
| DS0 | 64 Kbps | 1 DS-0 | E0 | 64 Kbps | 1E-0 |
| DS1 | 1.54 Mbps | 24 DS-0s | E1 | 2.048 Mbps | 32 E-0s |
| DS2 | 6.3 Mbps | 96 DS-0s | E2 | 8.45 Mbps | 128 E-0s |
| DS3 | 44.8 Mbps | 28 DS-1s | E3 | 34 Mbps | 16 E-1s |
| Not defined | | | E4 | 140 Mbps | 64 E-1s |

SONET, which was developed first, specifies a basic transmission rate of 51.84 Mbps, called the *synchronous transport signal 1* (STS-1). The equivalent optical signal is called **OC-1** (optical carrier 1). To ensure that both SONET and SDH match into a common multiplexing hierarchy, SDH defines its base level, the synchronous transport module 1 (STM-1), at 155.52 Mbps, which is three times the SONET base level. The optical line rates of both hierarchies are shown in Table 1–2.

**Table 1–2**   Interface Rates of the SONET/SDH Multiplexing Hierarchy

| SONET | BIT RATE | SDH |
|---|---|---|
| STS-1 / OC-1 | 51.84 Mbps | - |
| STS-3 / OC-3 | 155.52 Mbps | STM-1 |
| STS-12 / OC-12 | 622.08 Mbps | STM-4 |
| STS-24 / OC-24 | 1244.16 Mbps | - |
| STS-48 / OC-48 | 2488.32 Mbps | STM-16 |
| STS-192 / OC-192 | 9953.28 Mbps | STM-64 |

### Building a Complex SONET/SDH Carrier Network

A typical SONET/SDH network basically consists of four different network elements:

1. Add/Drop Multiplexer (ADM)
2. Terminal Multiplexer (TM)
3. Digital Cross-Connect (DXC)
4. Regenerator

All of these elements are interconnected using the service provider's fiber plant and form a typical SONET/SDH network, as shown in Figure 1–2.

ADMs used for ring networks and TMs used in linear topologies may be interconnected directly with dark fiber. If the distance between two multiplexers exceeds approximately 40 km, a regenerator must be placed in between the multiplexers. The regenerator ensures proper transmission through regenerating the optical signal, which has been degraded during optical transmission across the fiber.

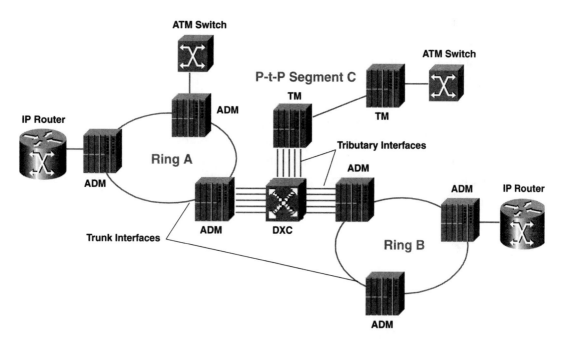

**Figure 1–2**    A typical SONET/SDH network consists of multiple multiplexers forming rings or linear links, which are interconnected with digital cross-connects

Multiplexers are equipped with two types of interfaces, trunk and tributary interfaces. ***Trunk interfaces*** are used to interconnect multiplexers. Trunk interfaces can range from OC-3/STM-1 to OC-192/STM-64, as specified in the current SONET/SDH recommendations. ***Tributary interfaces*** are used to attach client equipment, such as IP routers, ATM switches, or telephony switches, to the multiplexers. The range of available tributary interfaces for a multiplexer typically starts at the hierarchy level of the multiplexer's trunk interface and goes down to DS-0 or E1 interfaces.

To be able to switch a connection from a ring network segment to another or onto a point-to-point segment, DXCs are used. For example, in Figure 1–2, a DXC is placed between ring A, ring B, and the linear segment C. One multiplexer from each segment is attached to the DXC, using multiple tributary interfaces. Depending on what level of the TDM hierarchy connections have to be cross-connected, a certain type of DXC and the appropriate tributaries have to be used. If connections at DS1/E1 level are to be cross-connected, E1/DS1 tributaries must be used to attach the multiplexers to the DXC, and a DXC capable of switching at DS1/E1 level must be selected.

A network reference model is specified by the SONET/SDH standards to define a proper structure for all of the mechanisms in this complex network scenario. Three layers are defined (Figure 1–3). Each network element may be part of only one or more of these layers, according to its tasks and its requirements.

The lowest layer, called ***section layer*** (within SONET) or ***regenerator section layer*** (within SDH), incorporates optical transmission and signal regeneration. All network elements are part of this layer. Moreover, regenerators are only part of this layer, because they simply regenerate the optical signals along very long distance interconnections.

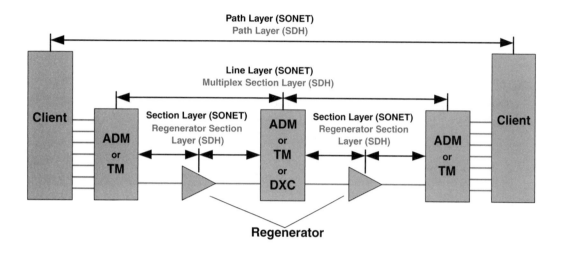

**Figure 1–3**    SONET/SDH network reference model introduces three functional layers

The second layer is called the *line layer* (SONET) or *multiplex section layer* (SDH). At this layer, several low-speed signals are multiplexed to and demultiplexed out of the high-bit-rate signal of the trunk interfaces. Multiplexers and DXCs are part of this layer.

The third layer is responsible for end-to-end connection delivery and is called the *path layer* (SONET and SDH). The client equipment terminating the endpoints of connections is part of this layer.

Certain control information is required at each layer for signaling, performance monitoring, or protection switching. This control information is carried in the *overhead* of each layer. Thus, a multiplexer is adding some path overhead (POH) to the signal coming from the tributary interface. It multiplexes multiple signals together and adds the section overhead (SONET) or multiplex section overhead (SDH). Finally, before sending it out to the next network element, it adds the line overhead (LOH) within SONET or the regenerator section overhead (RSOH) within SDH.

## Multiplexing and Framing Standards

A wide variety of signals can be transported across a SONET/SDH infrastructure. A signal inserted or extracted through a tributary interface is mapped into an appropriate container. This container is called *virtual tributary* (VT) within

SONET and *virtual container* (VC) within SDH. In addition to the signal itself, the container also includes the POH information, ensuring end-to-end connectivity.

How signals are multiplexed and mapped is specified in the SONET/SDH recommendations. The mapping and multiplexing scheme for the most important signal rates is outlined in Figure 1–4. Please keep in mind that this is not a complete picture of both the SONET and SDH multiplexing hierarchy.

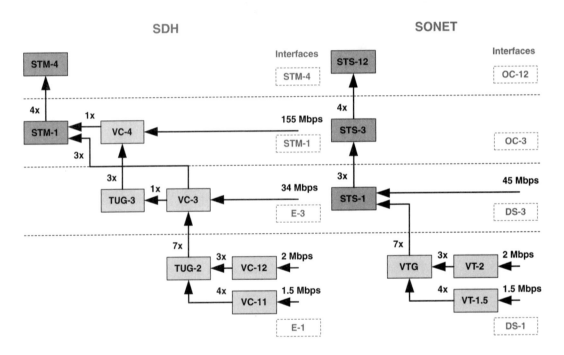

**Figure 1–4**    SONET and SDH multiplexing hierarchy

Within SONET, four VT1.5s and three VT2s are combined together into a VT group (VTG) at the DS1 level. Seven VTGs are then combined and mapped into the basic building block the STS-1, which resides at the DS3 level. Because the first optical interface specified (the OC-1 interface) is not commonly used in practice, the first optical SONET interface to be considered is the OC-3 interface running at 155 Mbps. Thus, after mapping the VTGs into the STS-1, three STS-1s are mapped into an OC-3 interface.

Within SDH, four VC-11s and three VC12s are combined together into a transport unit group (TUG-2) at the E1 level. Seven TUG-2s are then combined together and mapped into a VC-3 at the E3 level. As opposed to SONET, SDH also allows a high-bit-rate signal at 34 Mpbs to be mapped into a VC-3. The first specified optical interface within SDH is the STM-1 interface running at 155 Mbps (equivalent to the OC-3 interface of SONET). A VC-3 itself can be mapped in two different ways into an STM-1 interface. It can be mapped into a TUG-3, and three TUG-3s can be combined into a VC-4, which is then mapped into an STM-1 interface. The second possibility is to map three VC-3s directly into an STM-1 interface.

The basic building block of SONET is the STS-1, and the basic building block of SDH is the STM-1 frame. A high-level logical picture of the frame structure of both the STS-1 and STM-1 is shown in Figure 1–5. Keep in mind that the STS-1 frame is used for a 51.84-Mbps interface, and the STM-1 frame is used for a 155-Mpbs interface. Thus, the STM-1 frame has three times the columns of the STS-1 frame. From a functional point of view, the structures of the two frames are most likely the same.

The first columns of the frame are used for carrying the overhead of the first two SONET/SDH layers, the section overhead (SOH) and LOH, as defined in SONET, or the RSOH and the multiplex section overhead (MSOH), as defined in SDH. The remaining columns are used for the payload carrying the multiplexed containers. As already mentioned, the overhead of the third layer (the POH) is carried within each container.

The *payload pointers*, which are part of the LOH (SONET) or MSOH (SDH), are the key function in both synchronous hierarchies. The VTs and VCs possibly may not be synchronous with the STS/STM-Frame. Ideally, SONET/ SDH networks are synchronous, but in practice, there are always small clocking differences, and these have to be accommodated. As a consequence, the VTs and VCs are mapped into the STS/STM-frame at slightly varying boundaries. The payload pointers then are used to maintain a fixed relationship between the STS/STM-frame boundaries and the position of the containers mapped into the payload.

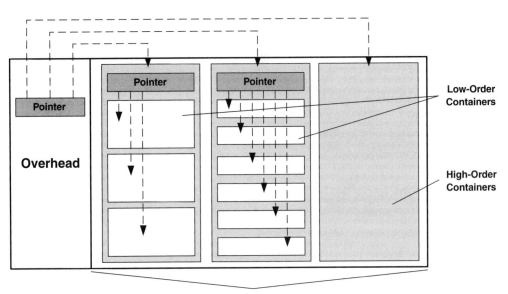

**Figure 1–5**    The SONET STS and SDH STM frame consist of an overhead and a payload portion

The pointers of the SOH/MSOH are referring to the position of the high-order containers in the frame. A high-order container can either carry a signal directly (a VC-3 may directly carry a 34-Mbps signal) or may carry low-order containers and have its own payload pointers, which point to the position of the low-order containers.

The second most important information carried in the SOH/MSOH is the K1/K2 overhead bytes that are used for protection switching purposes. Using the information carried in the K1/K2 bytes, a protection-signaling protocol is implemented and used by the SONET/SDH network elements to monitor the network status, detect failures, and react appropriately to restore failed traffic. Details on protection signaling and the protection architectures used in SONET/SDH networks are covered in Chapter 3, "Optical Networking Technology Fundamentals."

Much more information is carried in the LOH/RSOH and SOH/MSOH of the STS/STM-frame but the details are outside of the scope of this section. Additional information can be found in the SONET recommendation GR-253 [BELL-1] and the SDH recommendation G.707 [ITU-3].

### Providing Carrier Class Network Availability

Network availability, in particular, becomes a key design criteria in the case of providing telephony services, because these services usually require a nearly 100% uptime. The network must be designed to handle fiber cuts, as well as equipment failures. SONET/SDH has been specifically designed to fulfill that requirement.

In a SONET/SDH network, there is always some capacity reserved for protection purposes. Figure 1–6 illustrates that, in a typical ring architecture, there is always a working fiber (ring) and a protection fiber (ring). Usually in SONET/SDH, the protection path is not used under normal transport conditions; only in the case of a failure is the capacity of the protection path used to reroute and restore traffic affected by the failure. This rerouting is done solely based on the SONET/SDH network availability information and does not take into account any higher layer forwarding and reachability requests, i.e., ATM or IP reachability and availability information.

In the SONET/SDH frame overhead performance monitoring and protection information is exchanged. This information is used to implement a signaling protocol for protection switching, which is then used by the SONET/SDH network elements to communicate with each other and to perform required protection switching.

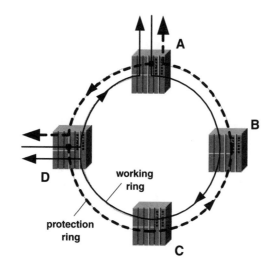

**Figure 1–6**    SONET/SDH provides comprehensive protection options

With the protection switching mechanisms implemented in the SONET/ SDH layer, failures typically can be restored within 60 ms; thus, the upper layers will usually not realize that a failure has occurred. It is important to understand that, whenever another technology or combination of technologies is going to replace SONET/SDH, all the protection mechanisms must be guaranteed by this new implementation. Otherwise, no service provider would replace an existing, reliable part of the network. This will be emphasized in the following chapters when we deal with replacement strategies in detail.

## Adding Versatility with Asynchronous Transfer Mode

SONET/SDH perfectly fulfills the needs of service providers that are mainly delivering telephony services. Old-world TDM telephony is connection-oriented and requires constant bandwidth (usually 64 Kbps per voice channel), even if no data has to be delivered (no speech has to be transferred via the connection). This is the reason why multiplexing and container allocation in SONET/SDH networks is optimized for voice. Static and fixed bandwidth is allocated permanently as soon as the desired connection is provisioned.

In the 1980s, standardization bodies driven by telcos and vendors developed a versatile layer on top of SONET/SDH to get a more flexible allocation of bandwidth, as well as more advanced service offerings. The new technology— ATM—works with synchronous slots called *cells*, but these cells do not have to be sent only periodically, as is the case with timeslots in TDM networks. Instead, there are also defined services sending cells on demand, for example, whenever new collected voice samples of a voice connection have to be transferred. This means that, during silence, no cells are sent and, thus, no bandwidth is blocked by idle cells.

Furthermore, numerous studies have shown that data services would play a more dominant part in the service portfolio of service providers. With this in mind and the fact that data traffic is more dynamic than static, it becomes obvious that static provisioning of timeslots or containers was no longer the most efficient and preferred method for carrier networks. As a consequence another important goal of ATM was to integrate voice and data traffic in a more efficient way than SONET/SDH.

An ATM network basically consists of a mesh of ATM switches interconnected by point-to-point ATM links or interfaces. The transmission or switching units of ATM are small, fixed-length data packets, so-called cells, with a fixed size of 53 bytes—thus the term *cell switching*. Because the cells have a small and fixed size, ATM switches can forward cells from one interface to another very quickly. In fact, it was one of the key design goals at the time to get a forwarding plane capable of switching information quickly (at interface speeds up to several Gbps) between different interfaces of ATM switches. Interestingly enough, in today's most advanced packet switches (routers that forward variable-length packets), the internal structure is also very often based on fixed-length cells, although these cells are, in most cases, of different size than 53-byte ATM cells.

ATM allows carriers to transport traffic via end-to-end connections across ATM networks with a predictable and very small delay. This is the key enabler for delivering voice and data services with a certain QoS in terms of available bandwidth, delay, and jitter.

ATM switches are typically interconnected with optical interfaces to the SONET/SDH hierarchy. The physical specifications have been taken from the SONET/SDH recommendations to ensure interoperability and the capability to directly connect ATM switches to SONET/SDH network elements.

## ATM Reference Model

The ATM architecture uses a logical model to describe the functionality it supports. ATM functionality corresponds to the physical layer and part of the data link layer of the Open Systems Interconnection (OSI) reference model.

The ATM reference model is composed of three planes that span all layers. The *control plane* is responsible for generating and managing signaling requests. The *user plane* is responsible for managing the transfer of data. Finally, there is the *management plane*, which consists of two components:

1. Layer management: This component manages layer-specific functions, such as detection of failures and protocol problems.
2. Plane management: This component manages and coordinates functions related to the complete system.

The ATM reference model is composed of the following ATM layers:

- The *physical layer* is analogous to the physical layer of the OSI reference model. The ATM physical layer manages the medium-dependent transmission.
- The *ATM layer* is tightly aligned with the ATM adaptation layer (AAL). The ATM layer is roughly analogous to the data link layer of the OSI reference model. The ATM layer is responsible for establishing connections and passing cells through the ATM network. To do this, it uses information in the header of each ATM cell.
- The *AAL,* which is strongly aligned with the ATM layer, is also roughly analogous to the data link layer of the OSI model. The AAL is responsible for isolating higher layer protocols from the details of the ATM processes.
- Finally, the higher layers residing above the AAL accept user packet user data and hand it to the AAL. Figure 1–7 shows the ATM reference model.

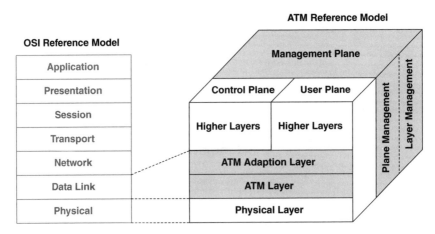

**Figure 1–7**    The ATM reference model includes three planes and three layers

## Introducing Statistical Multiplexing

Instead of supporting pure TDM, the interfaces of ATM switches can use the payload of the SONET/SDH frame as one big container. This option is called *virtual concatenation* within SONET/SDH. For example, a 622-Mbps ATM interface matches the transmission rate of an STM-4 interface in the SDH world. The payload is used as one big container. That means that the four VC-4s are glued together; the big container is called a ***VC-4c.***

To be able to distinguish between a SONET/SDH interface in standard TDM mode or in concatenated mode, a "c" is appended to indicate concatenation. For example, an OC-3c is a 155-Mbps SONET interface that can be used with ATM switches. An OC-3 SONET interface cannot be used with ATM switches but is used instead for standard SONET applications. By using the entire interface bandwidth as one big transport unit, the ATM cells can be dynamically mapped into it. This method is also referred to as ***statistical multiplexing*** and is much more efficient for transporting dynamic and elastic data traffic than is fixed TDM.

On top of the SONET/SDH network, a meshed ATM network can now be set up (Figure 1–8). The ATM switches are interconnected by concatenated SONET/SDH interfaces, for example, STS-12c/STM-4c or others.

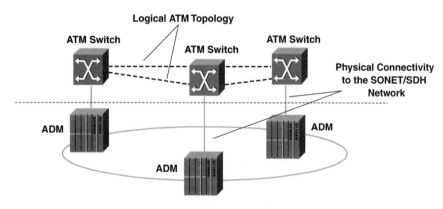

**Figure 1–8**    The ATM network is built on top of the SONET/SDH network to introduce statistical multiplexing

## Transport Multiple Services across a Single Network

The ATM adaption layer (AAL) is responsible for transforming the client signal into ATM cells. To be able to transport different types of traffic, such as voice, mission-critical data, and Internet traffic, different types of AAL have been defined.

The AAL translates between larger service data units (SDUs; for example, video streams and data packets) of upper-layer processes and ATM cells. Specifically, the AAL receives packets from upper-level protocols and breaks them into the 48-byte segments that form the Payload field of the ATM cell.

Several AALs are specified to support a wide range of traffic requirements. Table 1–3 shows the AAL types and their characteristics.

**Table 1–3**    ATM Incorporates Four Different Types of ATM Adaptation Layers with Certain Characteristics

| CHARACTERISTICS | AAL1 | AAL2 | AAL3/4 | AAL5 |
|---|---|---|---|---|
| Requires timing between source and destination | Yes | Yes | No | No |
| Bit rate | Constant | Variable | Available | Available |
| Connection mode | Connection-oriented | Connection-oriented | Connection-oriented or connectionless | Connection-oriented |
| Traffic types | Voice and circuit emulation | Packet voice and video | Data | Data |

The AAL is again broken down into two sublayers, the ***convergence sublayer (CS)*** and the ***segmentation and reassembly sublayer (SAR)***. The CS adapts information into multiples of octets. Padding can be added to achieve this purpose. The SAR segments information into 48-octet units at the source and reassembles them at the destination.

### AAL1

AAL1, a connection-oriented service, is suitable for handling circuit emulation applications, such as voice and video conferencing. Circuit emulation services also allow the attachment of equipment currently using leased lines to an ATM backbone network. AAL1 requires timing synchronization between the source and destination. For this reason, AAL1 depends on a media that supports clocking, such as SONET. The AAL1 process prepares a cell for transmission in the following way, also shown in Figure 1–9.

Synchronous samples (for example, 1 byte of data at a sampling rate of 125 μseconds) are inserted in the Payload field. Sequence Number (SN) and Sequence Number Protection (SNP) fields are added to provide information that the receiving AAL1 uses to verify that it has received cells in the correct order. The remainder of the Payload field is filled with enough single bytes to equal 48 bytes.

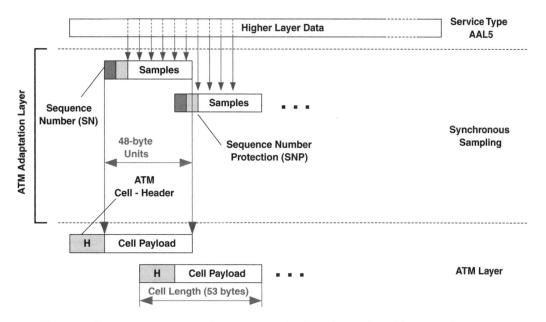

**Figure 1–9**   AAL1 maps synchronous samples into the 53-byte big ATM cells

### AAL2

AAL2 is optimized for the transport of variable-bitrate, time-dependent traffic over ATM. As with AAL1, the SAR provides for a sequence number, but in addition, a given cell will contain a designator as to whether it is the beginning, middle, or end of a higher layer information frame.

Appended to the SAR SDU is a length indication (LI) and a cyclic redundancy check (CRC). The resultant SAR-protocol data unit (PDU) is delivered to the ATM layer, where it forms the payload (ATM-SDU) of the ATM cell (ATM-PDU). An AAL2 cell can also be partially filled; in this case, unused bytes are padded.

The most important use for AAL2 is transfer of video, although AAL5 got a lot more momentum for this application. With some vendors implementing variable bit rate (VBR) over AAL1 for variable-rate voice, AAL2 may no longer be needed or used.

### AAL3/4

AAL3/4 supports both connection-oriented and connectionless data. It was designed for network service providers and is closely aligned with Switched Multimegabit Data Service (SMDS). AAL3/4 will be used to transmit SMDS packets over an ATM network. AAL3/4 prepares a cell for transmission in the following way, also shown in Figure 1–10.

The CS creates a PDU by prepending a Beginning/End Tag header to the frame and appending a Length field as a trailer.

The SAR fragments the PDU and appends a header to it. The SAR sublayer also appends a CRC-10 trailer to each PDU fragment for error control. The completed SAR PDU becomes the Payload field of an ATM cell, to which the ATM layer appends the standard ATM header.

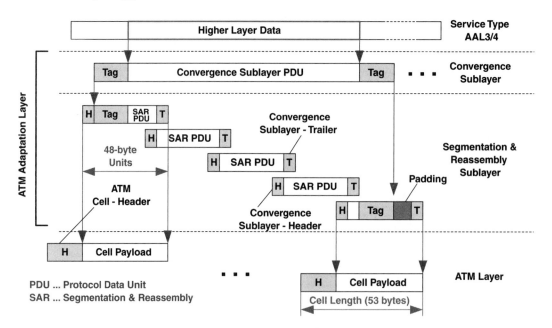

**Figure 1–10**    The AAL 3/4 incorporates two sublayers, the CS and the SAR

### AAL5

AAL5 is the primary AAL for data and supports both connection-oriented and connectionless data. It is used to transfer most non-SMDS data, such as Classical IP (CLIP) over ATM and LAN emulation (LANE). AAL5 is also known as the ***simple and efficient adaptation layer*** (SEAL), because the SAR sublayer merely accepts the CS-PDU and segments it into 48-octet SAR-PDUs, without adding any additional fields.

AAL5 prepares a cell for transmission in the following ways, also shown in Figure 1–11.

- The CS sublayer appends a variable-length pad and an 8-byte trailer to a frame. The pad ensures that the resulting PDU falls on the 48-byte boundary of an ATM cell. The trailer includes the length of the frame and a 32-bit cyclic redundancy check (CRC) computed across the entire PDU. This allows the AAL5 receiving process to detect bit errors, lost cells, or cells that are out of sequence.
- The SAR sublayer segments the CS PDU into 48-byte blocks. A header and trailer are not added (as in AAL3/4), so messages cannot be interleaved.
- The ATM layer places each block into the payload field of an ATM cell. For all cells except the last one, a bit in the payload type (PT) field is set to zero indicating that the cell is not the last cell in a series that represents a single frame. For the last cell, the bit in the PT field is set to one.

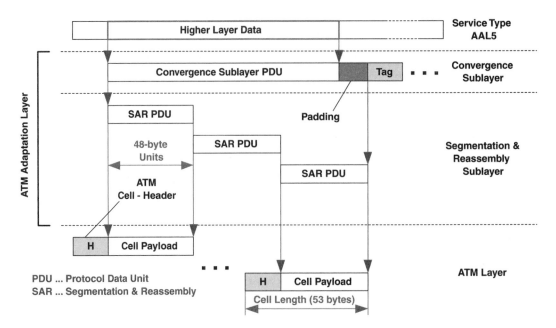

**Figure 1–11**    The AAL5 also incorporates two sublayers, the CS and the SAR

## ATM Services

ATM is a connection-oriented networking technology. As we will see throughout the following description, connection-oriented networking differs significantly from connectionless networking. One can best explain the differences by thinking of normal life examples.

Connectionless networking means that one sends data into a network, and the network knows itself by analyzing certain parts of the data (in IP networks, for example, the destination IP network address) where to forward the data. This process is also generally known as *routing*. Today's Internet is, to a large extent, built on that approach. For connectionless networking, the reader should be also aware that each data unit (in IP, generally called *IP packet*) can take a different path through the network, whereas, as we will see in connection-oriented networking, all data always travels the same path.

Connection-oriented networking is very much the same as we know from the good old telephone service. If you want to place a call to a remote user, you pick up the phone (initiate connection setup), then dial the number (define the destination of the remote user where the call should be set up), and, finally, let the telephone network establish the connection. After this successful setup, both parties can talk (exchange data in connection-oriented networking—often referred to as the ***data transfer phase***), and after finishing the conversation, both parties hang up (they tear down the connection). So we have three phases: connection setup, data transfer, and connection teardown.

End-to-end connectivity is achieved by provisioning virtual connections through the ATM network. Each ATM switch in the meshed network along the path of a virtual connection is configured to switch the virtual connection from the incoming interface to the right outgoing interface.

There are three types of ATM services:

**1.** Permanent virtual connection (PVC)

**2.** Switched virtual connection (SVC)

**3.** Connectionless service (which is similar to SMDS)

#### PERMANENT VIRTUAL CONNECTION

A PVC allows direct connectivity between sites. In this way, a PVC is similar to a leased line.

Advantages of PVCs are the guaranteed availability of a connection and the fact that no call setup procedures are required between switches. Disadvantages of PVCs include static connectivity and the manual administration required to set up.

#### SWITCHED VIRTUAL CONNECTION

An SVC is created and released dynamically and remains in use only as long as data is being transferred. In this sense, it is similar to a telephone call. Dynamic call control requires a signaling protocol between the ATM endpoint and the ATM switch.

Advantages of SVC include connection flexibility and call setup that can be automatically handled by a networking device. Disadvantages include the extra time and overhead required to set up the connection.

### ATM Virtual Connections

ATM networks are fundamentally connection-oriented. This means that a virtual channel (VC) needs to be set up across the ATM network prior to any data transfer. (A virtual channel is roughly equivalent to a virtual circuit.)

ATM connections are of two types:

**1.** Virtual paths, identified by virtual path identifiers (VPIs)

**2.** Virtual channels, identified by the combination of a VPI and a virtual channel identifier (VCI)

A virtual path is a bundle of virtual channels, all of which are switched transparently across the ATM network on the basis of the common VPI. All VCIs and VPIs, however, have only local significance across a particular link and are remapped, as appropriate, at each switch.

A transmission path is a bundle of VPs. Figure 1–12 shows how VCs concatenate to create VPs, which, in turn, concatenate to create a transmission path.

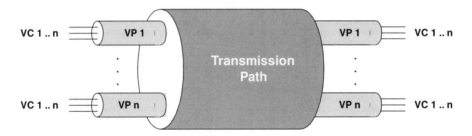

**Figure 1–12**    A typical ATM transmission path includes numerous virtual paths, which themselves includes multiple virtual circuits

## ATM Switching Operation

The basic operation of an ATM switch is straightforward. A cell is received across a link on a known VCI or VPI value. The connection value is then looked up in a local translation table to determine the outgoing port (or ports) of the connection and the new VPI/VCI value of the connection on that link. Then the cell is transmitted on that outgoing link with the appropriate connection identifiers. Because all VCIs and VPIs have only local significance across a particular link, these values get remapped, as necessary, at each switch.

## ATM Addressing

The ITU-T standardized upon the use of telephone-number-like E.164 addresses for public ATM (B-ISDN) networks.

The ATM Forum extended ATM addressing to include private networks. It decided on the "subnetwork" or "overlay" model of addressing, in which the ATM layer is responsible for mapping network layer addresses to ATM addresses. This model is an alternative to using network layer protocol addresses (such as IP and IPX) and existing routing protocols (IGRP, RIP).

The ATM Forum defined an address format based on the structure of the OSI network service access point (NSAP) addresses.

### SUBNETWORK MODEL OF ADDRESSING
The subnetwork model of addressing decouples the ATM layer from any existing higher layer protocols, such as IP or IPX. As such, it requires an entirely new addressing scheme and routing protocol.

All ATM systems need to be assigned an ATM address, in addition to any higher layer protocol addresses. This requires an ATM address resolution protocol (ATM_ARP) to map higher layer addresses to their corresponding ATM address.

### NSAP-FORMAT ATM ADDRESSES
The 20-byte NSAP-format ATM addresses are designed for use within private ATM networks, whereas public networks typically use E.164 addresses that are formatted as defined by ITU-T. The ATM Forum did specify an NSAP encoding for E.164 addresses. This will be used for encoding E.164 addresses within private networks but can also be used by some private networks. Such private networks can base their own (NSAP format) addressing on the E.164 address of the public user-to-network interface (UNI) to which they are connected and take the address prefix from the E.164 number, identifying local nodes by the lower order bits.

### ATM ADDRESS COMMON COMPONENTS
All NSAP-format ATM address-es consist of three components:

1. *Authority and format identifier (AFI)*: The AFI identifies the type and format of the initial domain identifier (IDI).
2. *Initial domain identifier (IDI)*: The IDI identifies the address allocation and administrative authority.
3. *Domain-specific part (DSP)*: The DSP contains actual routing information.

There are three formats of private ATM addressing that differ by the nature of the AFI and IDI:

1. *NSAP-encoded E.164 format*: In this case, the IDI is an E.164 number.

**2. DCC format**: In this case, the IDI is a data country code (DCC). These identify particular countries, as specified in ISO 3166. Such addresses are administered by the ISO National Member Body in each country.

**3. ICD format**: In this case, the IDI is an international code designator (ICD). These are allocated by the ISO 6523 registration authority (the British Standards Institute). ICD codes identify particular international organizations.

The ATM Forum recommends that organizations or private network service providers use either the DCC or ICD formats to form their own numbering plan.

Figure 1–13 shows the three formats of ATM addresses for use in private networks.

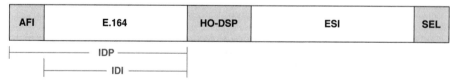

**Figure 1–13**    There are three address formats used in ATM networks: DCC, ICD, and NSAP addresses

## ATM Quality of Service

ATM supports QoS guarantees comprising the following components: traffic contract, traffic shaping, and traffic policing.

When an ATM end-system connects to an ATM network, it is making a "contract" with the network based on QoS parameters. This ***traffic contract*** specifies an envelope that describes the intended data flow. This envelope specifies values for peak bandwidth, average sustained bandwidth, burst size, and more.

ATM devices are responsible for adhering to the contract by means of ***traffic shaping***. Traffic shaping is the use of queues to constrain data bursts, limit peak data rate, and smooth jitter so that traffic will fit within the promised envelope.

ATM switches can use ***traffic policing*** to enforce the contract. The switch can measure the actual traffic flow and compare it with the agreed-upon traffic envelope. If the switch finds that traffic is outside of the agreed-upon parameters, it can set the cell loss priority (CLP) bit of the offending cells. Setting the CLP bit makes the cell "discard eligible," which means that any switch handling the cell is allowed to drop the cell during periods of congestion.

### ATM Signaling and Connection Establishment Overview

ATM signaling is a set of protocols used for call/connection establishment and clearing over ATM interfaces. The interfaces of interest to the ATM Forum are illustrated in Figure 1–14.

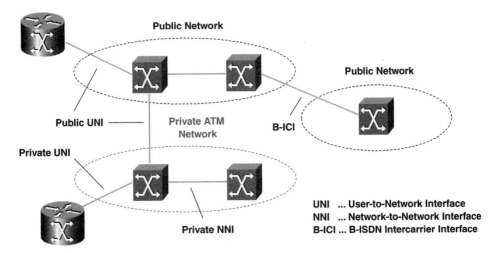

**Figure 1–14**   There are several types of interfaces specified by the ATM Forum

The public UNI is the user-to-network interface between an ATM user and a public ATM network. The private UNI is the user-to-network interface between an ATM user and a private ATM network. B-ISDN intercarrier interface (B-ICI) is the network-to-network interface between two public networks or switching systems. Private-to-network node interface (NNI) is the network-to-network interface between two private networks or switching systems.

When an ATM device wants to establish a connection with another ATM device, it sends a signaling request packet to its directly connected ATM switch. This request contains the ATM address of the desired ATM endpoint, as well as any QoS parameters required for the connection.

ATM signaling protocols vary by the type of ATM link. *UNI signaling* is used between an ATM end-system and ATM switch across ATM UNI. *NNI signaling* is used across NNI links.

### SIGNALING STANDARDS

The ATM Forum UNI 3.1 and UNI 4.0 specifications are the current standards for ATM UNI signaling. The UNI 3.1 specification is based on the Q.2931 public network signaling protocol developed by the ITU-T. UNI signaling requests are carried in a well-known default connection: VPI = 0, VPI = 5. For further reading, see the ATM Forum specification "ATM User to Network Interface Specification V3.1" [AF-2] or the ATM Forum Specification "UNI Signaling 4.0" [AF-3].

### ATM CONNECTION ESTABLISHMENT PROCEDURE

ATM signaling uses the "one-pass" method of connection setup used in all modern telecommunication networks (such as the telephone network). The ATM connection setup procedure works as follows. The source end-system sends a connection signaling request. The connection request is propagated through the network and reaches the final destination. The destination either accepts or rejects the connection request and, in case of accept, the connections get set up through the network. Figure 1–15 highlights the "one-pass" method of ATM connection establishment.

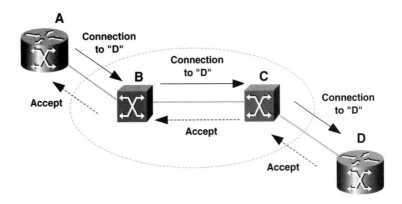

**Figure 1–15**    During the connection establishment, a request from the source end-system is propagated through the ATM network and either accepted or rejected by the destination end-system

CONNECTION REQUEST ROUTING AND NEGOTIATION
Routing of the connection request is governed by an ATM routing protocol. ATM routing protocols route connections based on destination and source addresses, and the traffic and QoS parameters requested by the source end-system. These routing protocols are usually either proprietary ones or the ATM Forum standard Private NNI routing protocol (PNNI).

The goal of the ATM Forum was to define NNI protocols for use within private ATM networks or, more specifically, within networks that use NSAP format addresses. Meanwhile, public networks that use E.164 numbers for addressing are interconnected using a different NNI protocol stack, based on the ITU-T B-ISDN user part (B-ISUP) ISUP signaling protocol. Negotiating a connection request that is rejected by the destination is limited because call routing is based on parameters of initial connection, and changing parameters might, in turn, affect the connection routing

## Providing Internet Services at the IP Layer

### Introducing the IP Layer on Top of the ATM Layer

For IP service delivery, the network layer and higher layers are required. Thus, IP routers are usually connected to ATM switches, which are interconnected across the ATM network by provisioning ATM virtual connections, as shown in Figure 1–16. These connections can be either static (where they are called *per-*

*manent virtual connections*, or PVCs) or dynamic (where they are called *switched virtual connections*, or SVCs).

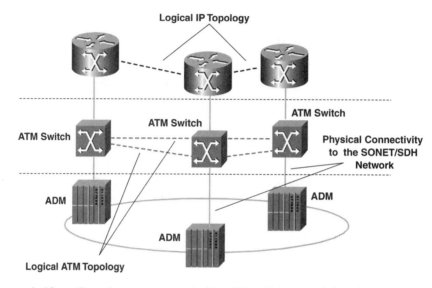

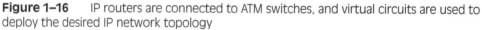

**Figure 1–16**    IP routers are connected to ATM switches, and virtual circuits are used to deploy the desired IP network topology

There are several methods used for running IP traffic over ATM networks. The first solutions used CLIP, defined by RFC 1577 [IETF-30].

In this case, the routers use an address resolution server—called the ***ARP server***—to find out the ATM addresses to connect to in order to reach an IP destination network. This approach is commonly known as the ***overlay model***, because the routers connected via an ATM cloud have no glue and, furthermore, cannot influence how the ATM network will set up a connection to reach the IP destination. There are several issues in terms of scalability and optimized routing throughout the ATM network, because CLIP works only in an IP subnet, and the ATM routing protocols have no information about the real needs of IP traffic delivery in terms of IP QoS, multicast, and others. Although CLIP was enhanced by additional protocols, such as Next Hop Resolution Protocol (NHRP) to overcome the limitation of suboptimal routing between subnets, the other issues (such as QoS mapping and IP-optimized connection setup) throughout the ATM network could not be solved.

Thus, the industry came up with other solutions by trying to combine the knowledge of IP destination reachability with the ATM backbones. These approaches are commonly known as ***peer-to-peer solutions***, because they can be viewed as running IP and ATM routing protocols on the same layer, making them real peers and making the ATM switches IP aware, thus enabling them to set up IP-optimized routes during connection establishment. There were several approaches, mostly driven by single vendors, such as IP switching, or TAG switching. Finally, MPLS became the standard way to go. It is currently already very stable because it is defined by the Internet Engineering Task Force (IETF) in several Internet standards (RFC) and will be discussed in the following paragraphs of this section.

## Using Multiprotocol Label Switching for Delivering IP Services

### SCALABILITY ISSUES OF STANDARD IP ROUTING PROTOCOLS

Normal IP routing is based on the exchange of network reachability information via a routing protocol, such as Open Shortest Path First (OSPF) or others. The routers are examining the IP destination address contained in the IP header of each incoming packet and are using the network reachability information to determine how to forward the packet. This process, also called ***routing lookup***, is independently performed at each router-hop along the path of a packet.

As described before, the IP network is typically deployed on top of an ATM infrastructure. This layer-2 network is completely independent from the IP network. This fact leads to two major scalability issues.

First, to provide optimal routing in the IP network, any-to-any connectivity between all routers is desired. This leads to the demand for $n*(n-1)/2$ virtual connections provisioned in the ATM network. Hence, whenever a router is to be added, a virtual connection has to be provisioned to each other router.

Second, network failures or topology changes provoke massive routing protocol traffic. Each router has to send routing updates across each virtual connection it is connected to in order to inform its neighbor about the new IP network reachability situation.

From the routing design point of view, another issue can be seen. As an example, consider a typical Internet service provider (ISP) network in which there are multiple routers at the edge of the network having peering relationships with other ISP routers to exchange routing information and to achieve global IP connectivity. To be able to find the best path to any destination outside the ISP's network, the routers in the core of the ISP network must also know all the network reachability information that the routers at the edge learn from the routers they are peering with. Hence, all core routers in the ISP network must maintain the entire Internet routing table, which requires a high amount of memory and leads to high CPU utilization. The solution to that issue is a forwarding technique where global network reachability is handled at the edge and packet forwarding "rules" are propagated to the core network.

### THE CONCEPT OF USING LABELS AS FORWARDING INFORMATION

MPLS intends to address the outlined issues. MPLS has been standardized within IETF over the past few years. MPLS introduces a fundamental new approach for deploying IP networks. It separates the control mechanism from the forwarding mechanism and introduces the "label" used for packet forwarding.

MPLS can be deployed in router-only networks to address the routing protocol design issues or can also be deployed in ATM environments for integrating both the layer-2 and layer-3 infrastructure into an IP + ATM network. An MPLS network consists of label switch routers (LSRs) in the core of the network and edge label switch routers (Edge-LSRs) surrounding the LSRs, as shown in Figure 1–17.

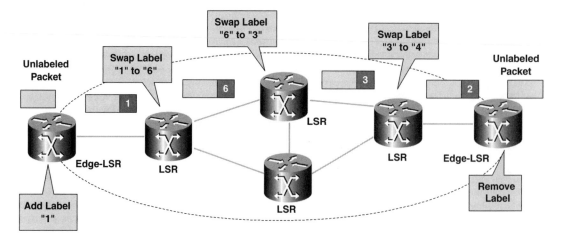

**Figure 1–17**   Edge-LSRs are imposing labels at the ingress side that are then used by the LSRs in the core to forward the traffic along the desired path and are removed again at the egress side by an Edge-LSR

Within the MPLS network, traffic is forwarded using labels. The Edge-LSRs at the ingress side of the MPLS cloud are responsible for assigning the label and forwarding the packets on to the next-hop LSR along the path that the traffic follows through the MPLS cloud. All LSRs along the path use the label as an index into a table that holds the next hop information and a new label. The old label is exchanged with the new label from the table and the packet is forwarded to the next hop. Using this method implies that the label value has only local significance between two LSRs. At the egress side of the network, the label is removed, and traffic is forwarded, using normal IP-routing protocol mechanisms.

MPLS introduces the term *forwarding equivalent class* (FEC). An FEC is a group of packets that share the same attributes while they travel through the MPLS domain. These attributes can be the same destination IP network, the same quality of service class, the same Virtual Private Network (VPN), etc. An FEC can also be a combination of some of these attributes. A label is assigned to each FEC (Figure 1–18), so all packets belonging to the same FEC will get the same label assigned from the LSR.

**Figure 1–18**    Each FEC gets its own locally unique label assigned, which is construed of four fields

The MPLS architecture allows that multiple labels forming a *label stack* are assigned to a packet. This might be for traffic engineering purposes, where one label is used to represent a tunnel and the other label to represent the FEC. Another possibility might be that one label is used to denote a certain VPN identifier, and another label is the FEC used for forwarding traffic across the provider network. To be able to distinguish whether label stacking is used, the *bottom of stack* (S) bit is part of each label. It is set to one for the last entry in the label stack (i.e., for the bottom of the stack) and zero for all other label stack entries.

Each IP packet has a *time to live* (TTL) field to prevent packets from looping endlessly in a network. The TTL field maintains a counter that gradually decrements down to zero, at which point the packet is discarded. A TTL 8-bit field is also included in the MPLS label, mainly as a troubleshooting mechanism.

A 3-bit field is reserved for **_experimental use_**. These bits are already used by some applications to determine the traffic priority and are to be used for providing certain levels of QoS in the MPLS network.

The rest of the label is used for the **_label value_** itself. This 20-bit field carries the actual value of the label.

The control and forwarding mechanisms of each LSR and Edge-LSR are separated into a control and forwarding plane (Figure 1–19). Two information structures are part of the **_control plane_**. The first one is the IP table routing maintained through running an Interior Gateway Protocol (IGP), such as OSPF or others.

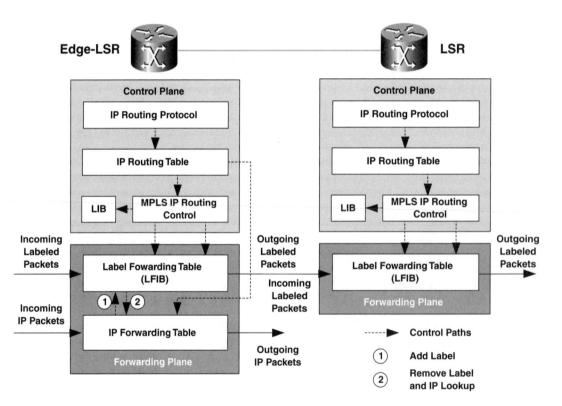

**Figure 1–19**   Each LSR (right side) or Edge-LSR (left side) has the control mechanism separated from the forwarding mechanism

The second one is a database called **_label information base_** (LIB). LSRs need to exchange the information on which label is associated with which FEC. There exist different protocols that perform this task. A protocol called **_Label_**

*Distribution Protocol* (LDP) is used to exchange the information on which prefix is associated with which label. To exchange the information on which VPN is associated with which label, the most common protocol is Multiprotocol BGP (MP-BGP). However, there is a variety of different protocols that enable the exchange of label-mapping information between LSRs.

Considering LDP as the protocol to be used, the label information is exchanged as follows. LDP uses **downstream label allocation**. This means that an LSR is told by its downstream neighbor which label to use when forwarding a packet belonging to a certain FEC out of the desired interface (Figure 1–20). Basically, there might be two situations. If the downstream LSR explicitly asks for a label binding for a certain FEC, and the downstream LSR answers with a label-binding message, this is called **downstream on demand**. If the upstream LSR did not ask for a specific label binding, and the downstream LSR sends out a label-binding message, this is called **unsolicited downstream** or simply **downstream label distribution**. Binding for the incoming label values is performed locally.

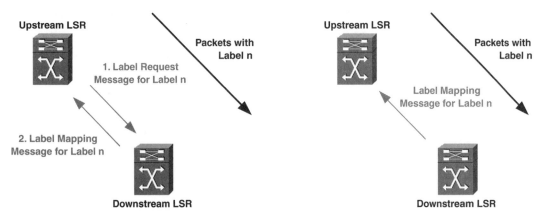

**Figure 1–20**    The LDP uses downstream label allocation, which might be requested by the upstream LSR (left half) or forced by the downstream LSR (right half)

The *forwarding plane* contains one information structure, which is a cache called *label-forwarding information base* (LFIB). The LFIB cache handles the actual packet-forwarding process. It contains information such as incoming label value, outgoing label value, prefix/FEC, outgoing interface (and encapsulation type of this interface), and the next hop address.

During the packet-forwarding process, this cache contains all required information to switch the packet or the cell through the LSR.

### MPLS-BASED APPLICATIONS

In the early drafts of MPLS, we could see the main application of MPLS in integrating IP and ATM. In the later drafts, we see that the authors are focusing more on MPLS applications such as VPNs and traffic engineering, because these concepts can be used as real differentiators between legacy-routed networks and MPLS networks. Service providers are trying to create and sell value-added services to their customers that they can use to differentiate themselves from the competition and to take a step up in the value chain. The opportunity for these service providers to offer VPN services based on IP, such as BGP/MPLS VPNs, does make this technology very attractive for the market. It also explains a little bit of the "hype" for MPLS that we could see in the market for the last two years. In our view, there is no slowdown of this situation visible on the horizon.

The most-deployed MPLS application is the MPLS VPN. A typical MPLS VPN network is shown in Figure 1–21. The customer access routers, also called *customer edge* (CE) routers, are connected to Edge-LSRs, acting as provider edge (PE) routers. The PE routers are assigning two labels to each packet. One label is representing the VPN identifier, and the top label is used to forward the packet through the network (Figure 1–21). The LSRs in the core of the network are called *provider* (P) routers and are performing standard label switching using the top label. The PE router at the egress side of the network is removing both labels and is using the second label to determine to which CE (VPN) the packets are to be forwarded.

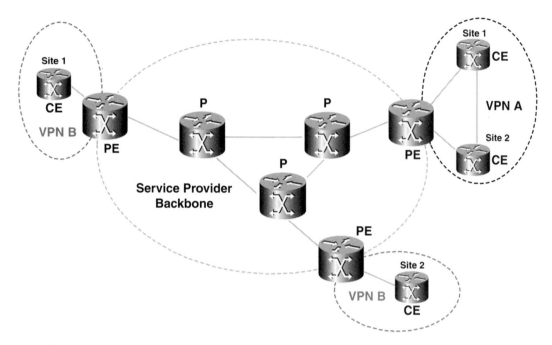

**Figure 1–21** MPLS VPNs—The PE routers use the label stack to determine to which VPN the routing information and traffic belongs

The second application makes use of the fact that the control plane is completely separated from the forwarding plane. Standard routing protocols compute the optimum path from a source to a certain destination, considering a routing metric such as hop-count, cost, or link bandwidth. As a result, one single least-cost path is chosen. Although there might be an alternate path, only the one determined by the routing protocol is used to carry traffic. This leads to an inefficient utilization of network resources.

MPLS Traffic Engineering (MPLS-TE) introduces the term *traffic trunk*, in which a group of traffic flows with the same requirements, such as resilience or traffic priority. MPLS-TE provides traffic-driven IGP route calculation functionality on a per-traffic-trunk basis. Other than with standard routing protocols, which are only topology driven, utilization of network resources and resilience attributes are analyzed and taken into account during path computation. As shown in Figure 1–22, multiple paths are possible from a source to a destination, and the best according to the current network situation is taken to ensure optimum network utilization.

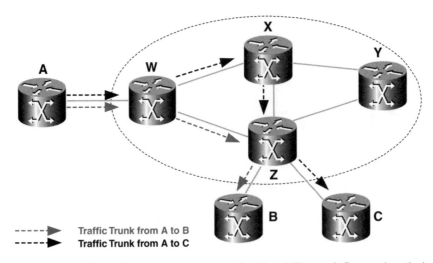

**Figure 1–22**    MPLS Traffic Engineering provides the ability to define and optimize the path to be taken by the traffic across the network

# Next-Generation Carrier Networks

The explosive growth of Internet/intranet traffic is making its mark on the existing transport infrastructure. An unprecedented shift has occurred in traffic content, pattern, and behavior. It has transformed the design of multiservice networks and has created a commercial demand for IP networks that operate in excess of 1 Gigabit per second. A change in usage patterns from fixed, configured, connection-oriented services to dynamic, connectionless IP services is currently evolving. According to several studies, telecommunication revenue will grow significantly in the near future, with data services, particularly IP, accounting for most of this increase. For public carriers, therefore, IP is critical for future revenue growth. Figure 1–23 shows how fast IP data traffic has been growing during the last few years and is expected to grow similarly in the next years.

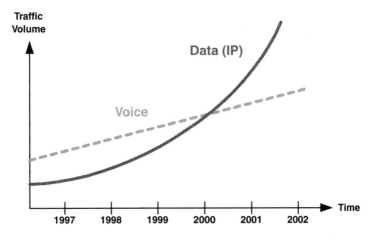

**Figure 1–23**    Data traffic has already overtaken voice traffic

As IP traffic volumes have exploded in recent years, bandwidth requirements for IP data have already reached limits that require the entire network architecture to be restructured in order to cope with the exponential growth.

IP backbone interconnections have already reached interface rates of OC-192/STM-64—thus, the highest level of SONET/SDH. Furthermore, the massive transport inefficiency of ATM, with its overhead up to 30%, overrides the advantages of the multiservice integration and QoS functionality.

As a consequence, the following trends and changes are forming the next-generation carrier networks.

## Eliminating Intermediate Layers in the Backbone

With IP as the dominant traffic in the network, the layered network architecture is no longer appropriate. The goal is to minimize the transmission overhead to maximize useful transport bandwidth. Certainly, complexity in provisioning, operations, network planning, and engineering must also be reduced to minimize the service provider's operational expenses (OPEX) and maximize the profit.

As can be seen from Figure 1–24, the SONET/SDH and ATM layers are going to be eliminated, transforming the backbone network into a two-layer network at the final stage. Fault detection and resolution, and network restoration become critical issues. By reducing the intermediate SONET/SDH layer, the comprehensive restoration mechanisms are removed and must be implemented within either the optical or IP layer. As shown in Figure 1–24, there are four network architectures possible. However, the question of the control plane for the optical network is important and not yet solved. Should the control be derived from the optical layer and run IP in an overlay fashion (notice the analogy to IP over ATM)? Should IP control the optical plane (peer model approach)? Should even other control planes deliver routing decisions (for example, content- or policy-derived control planes)? As stated earlier in this chapter, these issues will be discussed in greater detail in the Chapter 4, "Existing and Future Optical Control Planes."

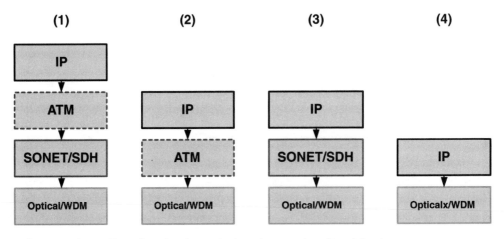

**Figure 1–24**    Transforming toward a two-layer network architecture

## IP over ATM over SONET/SDH over WDM (1)

Traditional service providers offering TDM, ATM, and IP services use networks consisting of all four layers: IP, ATM, SONET/SDH, and optical/WDM. The transmission efficiency is very low because of the high ATM overhead, including AAL5 and ATM cell overhead. High resilience against fiber cuts is provided through the SONET/SDH infrastructure but with a high initial equipment cost and high operational costs, due to complex provisioning and network management. WDM might be used to increase the capacity of the physical fiber plant by transmitting multiple wavelengths over a single fiber.

### IP over ATM over WDM (2)

A similar approach is to eliminate the SONET/SDH layer but to maintain the ATM layer. Service providers with a high amount of layer 2 services, such as leased lines and voice services, commonly use this approach and build a three-layer network, using the IP, ATM, and optical/WDM layer. The drawback of this architecture is, again, the inefficient transport of IP traffic because ATM is still there. On the other hand, all the other advantages of ATM, such as the strong QoS capabilities, remain.

### IP over SONET/SDH over WDM (3)

Alternative datacentric service providers today commonly implement a three-layer network architecture using the IP, SDH, and optical/WDM layer. Packet over SONET/SDH (POS) technology is used for transporting IP data directly over SONET/SDH infrastructure to eliminate the inefficient ATM transmission layer. SONET/SDH infrastructure might already be installed because it is in widespread use (especially in Europe) and can be used to provide protection functionality. It can also be used to transport legacy voice traffic.

### IP over WDM (4)

The fourth approach is to eliminate both the SONET/SDH and ATM layer, thus building only a two-layer network using the IP and optical/WDM layers. This approach is commonly known as an ***optical network*** and is mostly taken by service providers delivering data services and voiceover data service (VoIP).

The transport efficiency is optimized because the ATM layer is removed. The initial equipment is reduced because no SONET/SDH equipment is used. WDM technology may be used to increase the capacity of the fiber plant.

## Handling Fiber Capacity Exhaust

To meet the exponential growth in bandwidth demand of today's provider networks, Wavelength Division Multiplexing (WDM) technology is used to increase fiber capacity. Multiple wavelengths are sent over a single fiber. Each wavelength channel is capable of carrying any type of client signal. These client signals can be either ATM or SONET/SDH traffic but will mostly be IP traffic.

WDM systems typically have 4 or 16 channels. As transmission techniques evolve, systems with more than 40 channels have been developed and are called **Dense WDM** (DWDM) systems. ITU-T has standardized 100-GHz and 50-GHz channel spacings to ensure vendor interoperability. Closed WDM systems use a proprietary method to transport traffic over the WDM trunks and to provide standardized interfaces, such as SONET/SDH, ATM, POS, or Gigabit Ethernet, using wavelength transponders. Open WDM systems provide standard optical interfaces according to the ITU-T DWDM grid specifications and with SONET/SDH framing at 2.5 or 10 Gbps. These systems can be used with any SONET/SDH equipment, ATM switch, or IP router equipped with ITU-T wavelength grid conform interfaces.

WDM systems can be used with point-to-point connections but also in ring applications. Service providers have the ability to build up multigigabit connections in their networks. By using WDM technology, networks become very scalable. Bandwidth demands can be easily met by adding some wavelength channels to the system, providing additional "virtual fibers."

Furthermore, service providers can offer "optical leased lines" by providing dedicated wavelengths to customers. This revolutionary new type of service delivers enhanced flexibility to customers because of the bit rate independence of the wavelength service. Customers can build up traditional TDM-based networks with SONET/SDH or new-world data networks by directly interconnecting IP routers with different wavelengths.

## Adding Intelligence to the Optical Layer

Capacity is no longer the differentiating factor for service providers in the optical networking field. DWDM is commonly used and delivers massive amounts of raw bandwidth. Although DWDM allows multiple wavelengths to be transmitted over a single fiber, each wavelength must be terminated on an individual network element, typically a SONET/SDH ADM or IP router. This leads into an excessive amount of required equipment and a very complex and rarely manageable network architecture.

Transforming the optical layer from being a group of point-to-point pipes to a resilient, manageable optical network introduces devices called **wavelength routers**. Carriers can build a new optical layer that provides dynamic provisioning, reconfiguration for optimizing network resources, and protection and restoration at the wavelength level.

Wavelength routers are huge cross-connects, switching wavelengths through a meshed optical network (see Figure 1–25). Through the wavelength-routing capability implemented in the wavelength routers, end-to-end wavelength paths can be dynamically provisioned and restored. By managing the network at wavelength granularity, intelligent optical network elements scale to meet upcoming demands. Furthermore, carriers are able to assign different QoS levels or service types to wavelength paths.

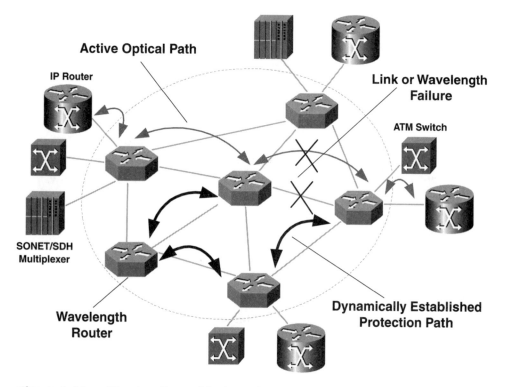

**Figure 1–25**    Wavelength provisioning using wavelength routers

Wavelength routers have a lot of additional benefits, such as rapidly increasing connectivity between service nodes, allocating bandwidth as dictated by service demands, and restoring connectivity between service nodes in event of failure. In addition, they allow providers to allocate restoration bandwidth according to business drivers, to simplify service network architectures, to offer next-generation bandwidth services with varying QoS at wavelength granularities, and to scale and introduce new technologies easily into their optical core.

A very important fact that must be mentioned is that the growth of service capacity will lead to higher costs if bandwidth is not manageable. It must be possible to manage the massive amount of new capacity provided by DWDM technology and to provision appropriate customer services. Basically, there are two approaches for managing a network. Centralized management ensures rapid provisioning; a single action propagates reactions throughout the network to activate a particular service or initiate rerouting across an alternative route. With distributed management, intelligence is put into every network element, and the network has "self-healing" capabilities. Therefore, network elements can react to failures without operator intervention and provide real-time survivability and reduced operational costs.

## Summary

In this chapter, we presented an overview of how traditional carrier networks have been designed until today. These designs have resulted in a complex network architecture, consisting of multiple network layers, where WDM provides optical transmission capacity, SONET/SDH delivers resilience against fiber cuts, ATM enhances the bandwidth utilization through statistical multiplexing and provides certain QoS functionality, and IP (on top of that architecture) delivers the interface to the service providers' data service offerings.

In contrast to these past implementations, next-generation carrier networks will be optimized for delivery of IP services while providing up to a terabit of capacity in the most scalable and flexible way. They will also provide easy, seamless, end-to-end provisioning. The key to this transition is to transform the old multilayer network into a two-layer network, meaning transporting IP directly on top of the optical layer. Having discussed these trends, we can now move on to the current and most popular standardization activities in this industry.

## Recommended Reading

[CIENA-1] CIENA Corp. Whitepaper, *Fundamentals of DWDM*.
[CIENA-2] CIENA Corp. Whitepaper, *The Evolution of DWDM*.

[CIENA-3] CIENA Corp. Whitepaper, *The New Economics of Optical Core Networks*, September 1999.

[CNET-1] CANARIE Inc., *Architectural and Engineering Issues for Building an Optical Internet*, draft, Bill St. Arnaud, September 1998.

[CSCO-1] Cisco Systems Inc. Whitepaper, *Scaling Optical Data Networks with Wavelength Routing*, 1999.

[CSCO-4] Networkers Conference—Session 606, *Advanced Optical Technology Concepts*, Vienna, October 1999.

[CSCO-5] Cisco Systems Inc. Whitepaper, *Cisco Optical Internetworking*, 1999.

[NFOEC-1] Technical Paper, *"IP over WDM" the Missing Link*, P. Bonenfant, A. Rodrigues-Moral, J. Manchester, Lucent Technologies, A. McGuire, BT Laboratories, September 1999.

[NFOEC-2] Technical Paper, *Packet over SONET and DWDM—Survey and Comparison*, D. O'Connor, Li Mo, E. Catovic, Fujitsu Network Communications Inc., September 1999.

[NFOEC-3] Technical Paper, *Position, Functions, Features and Enabling Technologies of Optical Cross-Connects in the Photonic Layer*, P. A. Perrier, Alcatel, September 1999.

# 2

# Optical Networking Standardization

**S**tandards are essential in any field, and this is especially true in the optical networking arena. Standards are required to define how to transport several service layer technologies across an optical transport infrastructure. The elimination of intermediate layers has led to the question of how Internet Protocol (IP), Asynchronous Transfer Mode (ATM), Synchronous Digital Hierarchy (SDH), and Frame Relay (FR) can commonly be mapped onto wavelengths provided by a Dense Wavelength Division Multiplexing (DWDM) network and how to impose special new requirements on the control planes for this optical transmission structure.

In this chapter, we will cover the most important standardization activities. These include the development of a uniform optical transport network (OTN) architecture, the specifications for optical interfaces up to 10-Gbps line rates, and the activities for defining a standard for dynamic control and wavelength provisioning in the OTN.

## Standardization Activities and Targets

An essential foundation for optical networking is the existence of standards that define how to transport the variety of service layer technologies across an optical transport infrastructure. The elimination of intermediate layers leads us to question how IP, ATM, SDH, and FR can commonly be mapped onto wavelengths provided by the DWDM network.

The first, and easiest, approach is to dedicate a separate wavelength for each different service. Because the number of available wavelengths might be limited, depending on the DWDM systems in use, this is not a very efficient method. A better solution is a common protocol stack acting as a control plane for both the service layer and the optical transport layer.

It should be kept in mind that SDH, ATM, and FR are being replaced more and more by IP. Therefore, the main focus of all the standardization activities centers on how to implement an IP-centric optical network.

This chapter covers the most important standardization activities, such as the development of a uniform (OTN) architecture, the specifications for optical interfaces at 10-Gbps line rates, and the activities for defining a standard for dynamic wavelength provisioning in the OTN.

# Activities of the T1 Committee and Its T1X1 Subcommittee

## Optical Networking Standardization Framework (G.871)

There is a tremendous demand for regulations in the fast-growing networking market to ensure vendor interoperability. Recommendations in the field of *optical networking (ON)* are particularly mandatory.

The *T1X1.5 working group* of the T1X1 subcommittee is working on a recommendation framework, called G.871, "Framework for Optical Networking Recommendations" [ITU-1], to control developments and new activities in ON. Table 2–1 shows which recommendations are currently under study and what issues they will discuss. In addition, the relevant recommendations for Synchronous Optical Network (SONET)/SDH and ATM are also listed to illustrate the relationships between the three networking areas.

**Table 2–1**    ITU-T Recommendations for ON, ATM, and SDH

| Topic | ON | SDH | ATM |
|---|---|---|---|
| Framework for Recommendations | **G.871** | | |
| Components & Subsystems | G.661, G.662, G.663, G.671, | G.661, G.662, G.663, G.671 | |
| Functional Characteristics | G.681, **G.798** | G.783, G.784, G.813, G.825, G.826, G.958, G.EPRMS | I.731, I.732 |
| Physical-Layer Aspects | **G.691, G.692, G.664, G.959.1** | G.703, G.957, G.691 | G.703, G.957, I.432 |
| Architectural Aspects | **G.872, G.873** | G.803, G.805 | G.805, I.326 |
| Structures & Mapping | G.709 | G.707, G.832 | I.361, I.362, I.363, I.610 |
| Management Aspects | **G.874, G.875** | G.774-x, G.784, G.831 | I.751 |

The recommendations in boldface are the new ON recommendations complementing already existing recommendations used in transmission networks.

The transmission aspects related to components and subsystems are already covered in G.671. The recommendation **G.664, "General Automatic Power Shutdown Procedures for Optical Transport Systems,"** complements this recommendation and covers automatic power shutdown procedures.

The recommendation **G.709, "Network Node Interface for the Optical Transport Network,"** covers all issues important for defining a standard node interface, such as framing, overhead functions, and payload mapping.

The functional aspects of optical network elements are described in the recommendation **G.798, "Characteristics of Optical Transport Network Equipment Functional Blocks."**

The most important recommendation is **G.872, "Architecture of Optical Transport Networks,"** which describes the layered structure of an OTN, including issues such as client/server layer associations, optical signal transmission, multiplexing, and routing.

The recommendation **G.873, "Optical Transport Network Requirements,"** covers WDM network design constraints and error/jitter performance necessary for supporting the wide variety of client signals.

How the network elements defined in G.798 can be managed is described in the recommendation **G.874, "Management Aspects of Optical Transport Network Element."** This recommendation is complemented by the recommendation **G.875, "OTN Management Information Model for the Network Element (NE) View,"** which defines the information required for managing the optical network elements defined in G.798.

The physical layer aspects for point-to-point WDM systems optimized for long-haul applications are covered in G.692. The new physical layer aspects for interoffice OTNs and metropolitan networks are covered in **G.959.1, "Optical Transport Network Physical Layer Interfaces."**

## OTN Architecture (G.872)

To define a unique architecture for OTNs, the recommendation G.872 has been developed. This recommendation uses the modeling methodology of recommendation G.805, "Generic Functional Architecture of Transport Networks." It assumes a layered structure and client/server layer associations. Each layer network is independent and serves specific functions, such as optical transmission, multiplexing, routing, or supervision.

The following sections are based on the proposed draft for recommendation G.872, "Architecture of Optical Transport Networks"[ITU-2].

### Layered Structure

An OTN relates to the physical layer of the International Organization for Standardization/Open Systems Interconnection (ISO/OSI) model, which is the first layer of the seven layers of this model. The OTN itself is defined as a *three-layer network*. The three layers are the optical channel (OCH) layer network, optical multiplex section layer network, and the optical transmission section (OTS) layer network.

Each layer network acts, on the one hand, as a server layer network for the layer network above (client) and, on the other hand, as a client layer network for the layer network below (server). The three-layer network is shown in Figure 2–1.

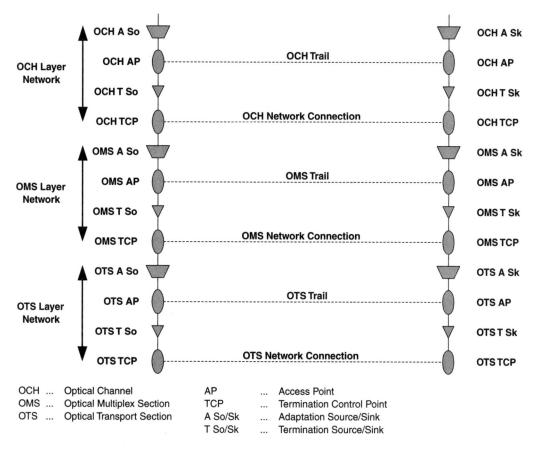

| | | |
|---|---|---|
| OCH ... | Optical Channel | AP ... Access Point |
| OMS ... | Optical Multiplex Section | TCP ... Termination Control Point |
| OTS ... | Optical Transport Section | A So/Sk ... Adaptation Source/Sink |
| | | T So/Sk ... Termination Source/Sink |

**Figure 2–1**   Layered structure of the OTN

### OCH LAYER NETWORK

The OCH layer network provides end-to-end connectivity networking for OCHs. There are several capabilities required for end-to-end networking.

First, it must be possible to rearrange OCH connections to provide flexible network routing. Second, the OCH overhead must be processed to ensure integrity of the OCH-adapted information. OCH supervisory functions are to be implemented to enable network level operations and management functions, such as connection provisioning, quality of service parameter exchange, and network survivability.

Because it is required by the heterogeneous service layer, OCHs can transparently carry optical client signals of several formats (SDH, ATM, Gigabit Ethernet, etc.). The optical client signal is passed through an OCH trail in an optical transmission unit (OTU). Each OTU has several properties, such as power level and signal-to-noise ratio (SNR) assigned. Furthermore, overhead information, including integrity check, is appended, and a modulation scheme is defined.

The OCH termination implements the supervisory functions, such as connectivity integrity check, loss of signal (LOS) detection, and alarm indicator signal (AIS) processing. The OCH termination source merges the OCH overhead and adapted client signal. The OCH termination sink accepts the received OTU and separates the client signal from the OCH overhead.

### OPTICAL MULTIPLEX SECTION (OMS) LAYER NETWORK

The OMS layer network provides necessary functions for usage of multiwavelength transmission systems (DWDM systems). Using multiple wavelengths increases the operation and management efficiency of optical networks.

The OMS layer network provides the required capabilities for the OMS connection rearrangement, which ensures flexible multiwavelength network routing. Furthermore, OMS overhead is processed to ensure integrity of the multiwavelength OMS adapted information. Finally, the OMS supervisory functions enable section-level operations and management functions, such as multiplex section connection provisioning and network survivability.

The OCHs are passed through an OMS trail. OCHs are combined to an OTU group of the order n (OTUG-n). The OMS termination source/sink performs supervisory functions and adds/removes OMS overhead similar to the OCH layer network.

### OPTICAL TRANSMISSION SECTION (OTS) LAYER NETWORK

The OTS layer network provides the functionality for transmission of optical signals over different media types, such as single- and multimode fiber (SMF and MMF). The multiplexed OCHs are transported in an optical transport module (OTM) with appropriate optical parameters (e.g., SNR and power level) over a defined wavelength range. The OTS termination performs supervisory functions, adds/removes OTS overhead, and inserts the optical supervisory channel for optical amplifier and repeater control.

### INTERLAYER ADAPTATION

Because traffic is transported across the OTN, each of the three layers has to be processed. Between each of these layers, interlayer adaptation has to occur.

### CLIENT SIGNAL/OCH ADAPTATION

The adaptation of a client signal to the OCH contains client-specific and server-specific processes. In general, client signal/OCH adaptation includes all processes required to generate a continuous data stream, which can be modulated onto an optical carrier or a wavelength, respectively.

Depending on the client signal digital mapping, payload scrambling and/or channel coding, such as nonreturn to zero (NRZ) coding, are applied. The output is a data stream with defined bit timing, framing information, and DC balance, which is handed over to the OMS layer. If a client signal already provides the appropriate data stream, the adaptation processes simply pass the signal to the OMS layer.

Client signal/OCH adaptation is one of the major points of discussion to provide interoperability between routers/switches and DWDM equipment from different vendors. A common encapsulation and framing (PPP over SONET/ SDH, ATM, Gigabit Ethernet, FR, etc.) with a defined line coding (NRZ, RZ, etc.) has to be considered to create a ***standard interface***.

### OCH/OMS ADAPTATION

OCH/OMS adaptation performs wavelength allocation for each OCH and modulates the channel's data stream onto the allocated optical carrier (wavelength). In the second stage, all channels are multiplexed together and handed over to the OTS layer.

### OMS/OTS ADAPTATION

OMS/OTS adaptation adds a supervisory channel to the optical multiplex entity for monitoring and management purposes.

# Activities of the Institute of Electrical and Electronics Engineering (IEEE)

## 10-Gbps Ethernet

The IEEE is currently working on two very important optical networking topics. The first one is the evolution of Ethernet technology to a transmission rate of 10 Gbps. With the development of Gigabit Ethernet some time ago, very inexpensive interfaces with a transmission rate at 1 Gbps have been made available. Compared to the expensive ATM and Packet over SONET/SDH (POS) interfaces, Gigabit Ethernet provided a very cost-effective solution for building up point-to-point Gigabit connections.

To accommodate the tremendous traffic growth, interfaces at 10 Gbps are required. As a consequence, the IEEE founded the 802.3ae study group to develop a standard 10-Gigabit Ethernet interface to provide the most inexpensive 10-Gbps solution for service providers. This standard is supposed to be finished in the first half of 2002.

The goal of this group is it to take the existing 802.3 media access control (MAC) layer, which has already been used for 10/100/1000-Mbps Ethernet. The interface will support only full-duplex operation, unlike the other interfaces of the Ethernet family, which also provide half-duplex operation.

To provide as much flexibility as possible at the physical layer for both equipment vendors and their customers, a media-independent interface (MII) is also to be specified, which supports the use of a certain type of already-installed cabling, together with appropriate transceivers. In addition to the MII, two physical interface specification approaches are basically followed in the study group. As a result, two different physical (PHY) specifications will be developed, in addition to the MII. The Local Area Network (LAN) PHY is targeted for 10-Gbps transmission across dark fiber and transparent WDM systems. The Wide Area Network (WAN) PHY will be compatible to the existing SONET/SDH infrastructure. The WAN PHY will operate at a data rate compatible to the payload rate of an OC-192c/STM-64c interface.

Optical transmission will be supported across both SMF and MMF. For MMF, the supported link distances will be up to 300 m. For SMF, multiple interface variations supporting link distances up to 2 km (short reach), 10 km (intermediate reach), and 40 km (long reach) will be available.

All activity areas of the 802.3ae group are shown in Figure 2–2 on the right side. To point out the similarities between the Gigabit and 10-Gigabit Ethernet specifications, the activities from the 802.3z group, which developed the Gigabit Ethernet specifications, are shown on the left side.

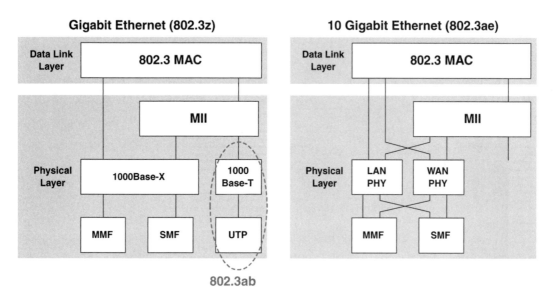

**Figure 2–2**    The 802.3ae group is adapting the work done by the 802.3z group for developing the 10-Gigabit Ethernet specifications

## Resilient Packet Rings

Fiberoptic rings are commonly deployed in metropolitan area networks (MANs) and also in WANs. Unfortunately, the available technologies used today with these rings do not fulfill the requirements of service providers.

SONET/SDH is very bandwidth-efficient for transporting IP traffic, as previously discussed. POS improves the efficiency but does not provide ring protection functionality to ensure high resilience in the ring.

As a consequence, the IEEE founded the 802.17 Resilient Packet Ring (RPR) working group in December 2000 to develop standards for RPR protocols used for the transport of data packets at data rates up to multiple Gigabits per second.

As the IEEE has already done in most cases, a single MAC layer algorithm for the data link layer and multiple physical layer options will be developed to provide as much flexibility as possible. Figure 2–3 is showing the elements of the data link and physical layer to be defined in the 802.17 specification.

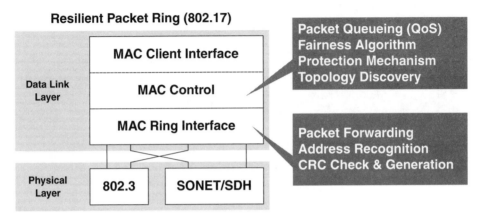

**Figure 2–3**    The 802.17 specification is defining enhanced control mechanisms at the MAC layer for protection and bandwidth management

As can be seen, at the data link layer, the RPR MAC layer might be split into three sublayers. In the middle, there is the MAC control sublayer, which incorporates all the enhanced functionality, such as bandwidth fairness and protection. Above that, there is the MAC client interface sublayer responsible for adapting the control primitives to the attached network protocol client. Below the MAC ring interface is the sublayer responsible for packet forwarding and addressing.

The layer-2 MAC protocol to be standardized will provide efficient bandwidth-sharing functionality using spatial reuse mechanisms in order to optimize the use of the available bandwidth in the ring. The specified RPR frame format will allow easy mapping of 802.3 frames into RPR frames and vice versa. Because data traffic is very elastic, dynamic and fair bandwidth allocation will be supported. To ensure correct resilience against failures (such as fiber cuts or interface failures), a protection mechanism will be implemented. All control mechanisms will scale to support rings up to 128 or 256 nodes.

At the physical layer, interfaces with transmission rates up to 10 Gigabits at a minimum will be standardized. The physical layer specifications will be defined in alignment with the 802.3 working group, International Telecommunication Union (ITU), and ANSI. Thus full-duplex Gigabit Ethernet and 10-Gigabit

Ethernet, as well as SONET/SDH physical interfaces typically ranging from OC-12c/STM-4c to OC-192c/STM-64c, will be available.

The SONET/SDH compliance at the physical layer is especially very interesting. It is important to be able to deploy RPRs on top of existing SONET/SDH infrastructures and on top of DWDM infrastructures requiring SONET/SDH framed client equipment interfaces.

Several equipment vendors, such as Cisco Systems, Nortel Networks, and Dynarc, already have developed their own proprietary solutions for RPRs. These solutions are analyzed and formed together into one standard solution formalized by the IEEE working group. Cisco Systems already submitted its Spatial Reuse Protocol (SRP), specified in the RFC2892 [IETF-8], to the 802.17 working group. Details on this solution can be found in Chapter 3, "Optical Networking Technology Fundamentals."

# Activities of the Optical Internetworking Forum (OIF)

## OIF Mission

In April 1998, the Optical Internetworking Forum (OIF)—an open forum focused on accelerating the deployment of optical internetworks—was founded by Cisco Systems, Inc., CIENA, AT&T, Bellcore, Hewlett-Packard Company, Qwest, Sprint and WorldCom. Since its founding, more than 100 companies representing the leaders of the telecommunications industry have joined the OIF.

The OIF provides a venue for equipment manufacturers, users, and service providers to work together to resolve issues and develop key specifications to ensure the interoperability of optical internetworks.

The goal of the OIF is to complement the activities and efforts of other standardization bodies, such as the ITU, IEEE, and others, which have already begun to standardize facets of the optical network layer. The OIF will identify opportunities in which early industry consensus can accelerate technology deployment and avoid a proliferation of vendor-specific approaches. Any specifications developed in this manner by the OIF will be provided as input to traditional standards bodies and other industry groups.

There are several areas that represent immediate opportunities to impact the evolution of optical internetworks. These areas include integrated management of all layers of an optical internetwork, data-optimized interfaces between data and optical equipment, as well as enhanced protection and restoration between network layers

The OIF has created four working groups, each responsible for a certain area, such as architecture, physical layer, signaling, and operation administration maintenance and provisioning (OAM&P). These four working groups are active in the working areas shown in Figure 2–4.

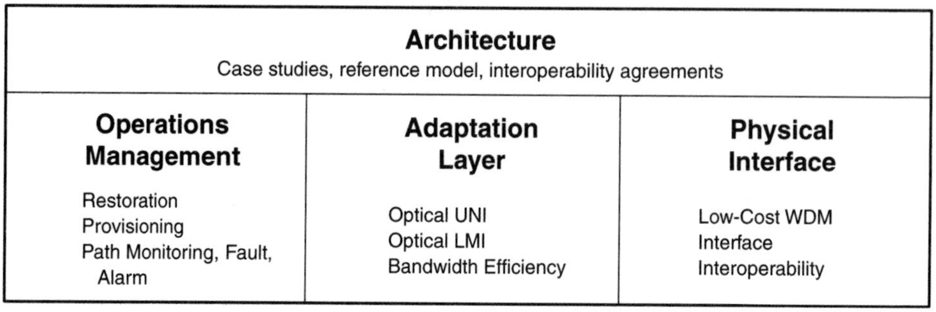

| Architecture | | |
|---|---|---|
| Case studies, reference model, interoperability agreements | | |
| **Operations Management** | **Adaptation Layer** | **Physical Interface** |
| Restoration Provisioning Path Monitoring, Fault, Alarm | Optical UNI Optical LMI Bandwidth Efficiency | Low-Cost WDM Interface Interoperability |

**Figure 2–4**    Optical Internetworking Forum working areas

The OIF's mission includes several goals. One, of course, is to achieve widespread industry participation. This ensures that as many equipment vendors and customers as possible try to agree on the standards to be proposed by the OIF. Besides this, the main task is to develop interoperability specifications for all the different network elements used when building an OTN. Uniform testing and validation criteria have to be specified to make common national and international standardization processes possible.

Besides the technical tasks, the OIF tries to highlight current and upcoming trends in the area of optical networking and strives to make people aware of what influence those have on their networks. By also participating in several networking events, the organization should be able to educate its audience.

According to the OIF kick-off meeting presentation [OIF-1],the relation to existing standardization bodies done by traditional organizations (ITU, Internet Engineering Task Force [IETF], etc.) is as shown in Figure 2–5.

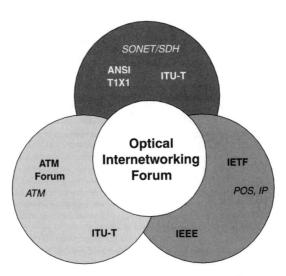

**Figure 2–5**    OIF in relation to existing standardization bodies

## Standard Physical Layer Interface

In January 1999, the physical layer working group proposed POS to the OIF as the physical layer for interfaces up to OC-192c/STM-64c with the OIF contribution "A Proposal to Use POS for the OIF Physical Layer up to OC-192c"[OIF-2].

In the last year, POS has been used to implement a wide variety of router interfaces in order to provide an efficient, high-speed data transport solution. There are already POS interfaces available with line rates starting at OC-3c/STM-1 up to OC-192c/STM-64c.

The characteristics of a POS interface perfectly fit into the requirements of a physical interface used in the OTN. First of all, it introduces very little overhead. The PPP encapsulation with byte-stuffed HDLC such as framing adds less than 1% overhead, on average. The provided reliability is very high at today's typical fiber error rates.

Furthermore, the $x^{43} + 1$ self-synchronous scrambler protects the SONET/SDH payload from malicious killer packet attacks. This ensures that interfaces extracting the clock rate from the line signal are not in danger to get out of sync.

Finally, one of the most important advantages is that router and DWDM system vendors already tested and optimized interoperability of their implementations.

The implementation of the POS physical layer is documented in a combination of IETF RFCs (RFC 2615, RFC 1661, and RFC 1662), along with Bellcore and ITU documents [BELL-1], [ITU-3].

For details on POS, refer to Chapter 3, "Optical Networking Technology Fundamentals."

## IP-Centric Control and Signaling for Optical Paths

The signaling working group of the OIF is responsible for developing a standard way for automated lightpath setup within an OTN.

In January 2000, the OIF contribution "IP Centric Control and Signaling for Optical Lightpaths" [OIF-3] proposed some concepts for provisioning lightpaths in the OTN. It is clear that IP currently is, and will be in the future, the dominating traffic type and, therefore, should also be used to control service provisioning in the OTN.

The Multiprotocol Label Switching (MPLS) architecture perfectly fits into the requirements of the OTN. The MPLS traffic engineering features must only be extended to support optical lightpath provisioning. Standard routing protocols, such as Open Shortest Path First (OSPF) and IS-IS (Intermediate System to Intermediate System protocol), are reused and adapted to support optical resource information distribution. As a signaling protocol, an extended version of Label Distribution Protocol (LDP) can be used. The advantage of this approach is that already-proven technology is used, and no 100% new protocols must be invented.

## Optical User-to-Network Interface (O-UNI)

Proprietary implementations providing dynamic service provisioning in an OTN are already available. End-to-end lightpaths can be established dynamically across an OTN using equipment from a single vendor. In addition to efforts in standardizing the dynamic establishment of lightpaths, described in the previous section, the OIF has already finalized the specification for a standard interface between an OTN, using a proprietary lightpath provisioning implementation, and the service layer network on top of it, using the provisioned lightpaths for building the desired network layer topology.

This standard O-UNI makes it possible to integrate a proprietary solution at the OTN with, for example, the MPLS control plane of the IP network at the service layer in an overlay model. In such a scenario, MPLS is used to establish data paths through the IP network. As the data path crosses the border of the OTN, the UNI signaling mechanism is used to translate the data path setup request into a lightpath setup request. A lightpath is then provisioned automatically across the OTN, and the data path can be transported through the OTN. The UNI 1.0 specification finalized by the OIF and proposed in the contribution OIF2000.125.3 "User Network Interface (UNI) 1.0 Signaling Specification" [OIF-4] concentrates on SONET/SDH-based interfaces. It covers signaling mechanisms for setting up connections at OC-3/STM-1 level or higher.

IP routers or ATM switches are connected to optical network elements (ONEs) with the UNI interface and are acting as clients to the OTN. The optical network elements placed in the OTN, which could be optical cross-connects (OXCs), wavelength routers, or advanced SONET/SDH systems, are interconnected with an optical network-to-network interface (O-NNI).

The UNI 1.0 specification defines three different types of UNI signaling reference configurations. In the first reference configuration, all routers or ATM switches connected to the OTN have integrated UNI *client-side signaling functionality* (UNI-C) and the network elements of the OTN have integrated UNI *network-side signaling functionality* (UNI-N). Figure 2–6 shows the resulting network model when using this UNI reference configuration to interconnect the service layer with the OTN layer.

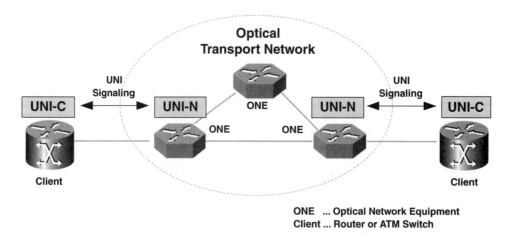

ONE   ... Optical Network Equipment
Client ... Router or ATM Switch

**Figure 2–6**    The proposed UNI 1.0 specification defines the interface between the network elements at the network and optical transport layer

The second reference configuration takes into account the fact that the optical network elements may not have integrated network-side signaling functionality. In this case, a separate signaling UNI-N agent is used for the OTN, which provides the UNI network-side signaling function.

The third reference configuration handles the case where both the optical network elements and the clients (router or ATM switches) do not have integrated signaling functionality. Therefore, a signaling UNI-N agent is used in the OTN for network-side signaling, and a UNI-C proxy is used to provide client-side signaling for one or more clients connected to the OTN.

Figure 2–7 shows both the second and third reference configurations. The OTN includes a UNI-N agent in its optical control plane to provide UNI network-side signaling to the attached clients. On the left side, a router acting as a client is providing UNI-C by itself, according to the second reference configuration. On the right side, a UNI-C proxy is providing the UNI-C for the right-hand side router, according to the third reference configuration.

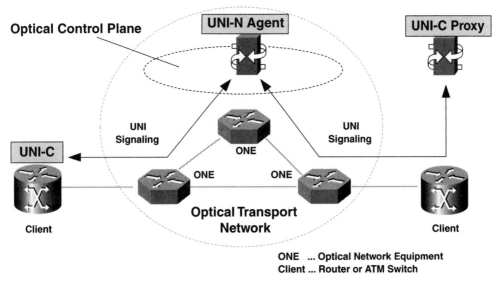

ONE  ... Optical Network Equipment
Client ... Router or ATM Switch

**Figure 2–7**    UNI signaling functionality might not be integrated in the network elements and, thus, could be provided by a UNI-N agent for the OTN or by a UNI-C proxy for the clients

A set of UNI signaling messages and attributes is defined. The UNI 1.0 signaling is based on the two MPLS signaling protocols, LDP and RSVP with Traffic Engineering Extensions (RSVP-TE). Both are used in MPLS networks to exchange the information necessary for each network element to know how to forward data across the network. So the UNI specification defines certain LDP and RSVP extensions to be able to carry UNI signaling information in LDP and RSVP-TE messages.

MPLS and its signaling protocols LDP and RSVP-TE are covered in the upcoming chapter, "Optical Network Technology Fundamentals." For further details, refer to the IETF drafts [IETF-6], [IETF-24], [IETF-25].

The UNI specification also specifies a *neighbor discovery* procedure that uses a Hello protocol. Thus, a ONE and a directly attached client device are able to discover each other automatically and verify the consistency of configured parameters.

A signaling channel is used to exchange control information between the two neighbors. This control channel can either be an in-band or an out-of-band signaling channel and is used to set up IP connectivity. The UNI specification defines how the ***IP signaling control channel*** is set up and maintained in various configurations.

Depending on what type of network is built on top of the OTN, different services are required. The UNI specification defines a group of service to be requested using UNI signaling. These are, for example, IP and ATM. Also, a service discovery procedure is proposed to allow a client device to determine the particulars of the UNI services offered and to negotiate certain parameters.

Details on how the optical UNI fits into the architecture of next-generation optical networks are covered in Chapter 4, "Existing and Future Optical Control Planes."

## Activities of the IETF

The IETF as the traditional standardization body of the Internet is also very active in the optical networking field. Some Internet drafts have already been proposed for defining a standard optical lightpath-provisioning framework.

The IETF follows an approach similar to the one from the OIF. MPLS seems not only to be an ideal architecture for integrated IP + ATM networks but also for the OTN. As a consequence, the IETF also proposes to use MPLS in the OTN and is on the way to specify the Multiprotocol Lambda Switching (MPLmS) architecture. A virtual circuit switched through an ATM network is comparable to a wavelength path (lightpath) switched through the OTN. Instead of a VPI/VCI combination the label is represented by a certain wavelength. OSPF and IS-IS are to be extended to support optical routing. RSVP or LDP are adapted to be able to set-up lightpaths through the network.

Details on this approach are covered in Chapter 4, "Existing and Future Optical Control Planes."

## Summary

In this chapter, we covered the most important standardization activities, which are necessary to create a unified OTN architecture ensuring the necessary vendor interoperability. Whereas T1X1 mostly focuses on optical transport technology to define standards for DWDM transmission technology, other standardization bodies focus more on the interface specifications between the OTN layer and the IP network layer.

The IEEE intensively works on standards for using Ethernet technology in OTNs as well, thus pushing Ethernet everywhere end-to-end. We see the most important arguments for that trend in the simplicity of Ethernet opposed to SONET/SDH and, of course, not to forget the low price of Ethernet-based interfaces, compared with old-world technology-based interfaces.

The goal of the OIF is to tie all standardization bodies together. With the UNI 1.0 specification, the OIF for the first time introduces a uniform optical control plane and allows proprietary wavelength provisioning implementations to be integrated with the IP or ATM network layer on top of it.

Having provided an overview of where the industry stands in terms of designing and deploying a future-proof network architecture facilitating IP services across a high-capacity optical transport infrastructure in the first two chapters, we can now start discussing related optical networking technologies in more detail in the next chapter

# Recommended Reading

[IETF-13] IETF-draft, draft-awduche-mpls-te-optical-01.txt, *Multi-Protocol Lambda Switching: Combining MPLS Traffic Engineering Control with Optical Crossconnects*, work in progress, November 1999.

[IETF-14] IETF-draft, draft-basak-mpls-oxc-issues-01.txt, *Multi-Protocol Lambda Switching: Issues in Combining MPLS Traffic Engineering Control with Optical Crossconnects*, work in progress, November 1999.

[ITU-1] ITU-T Draft Recommendation G.871, *Framework for Optical Networking Recommendations*, October 1998.

[ITU-2] ITU-T Draft Recommendation G.872, *Architecture of Optical Transport Networks*, July 1998.

[ITU-3] ITU-T Recommendation G.707, *Network Node Interface for the Synchronous Digital Hierarchy (SDH)*, March 1996.

[ITU-4] ITU-T Recommendation G.692, *Optical Interfaces for Multichannel Systems with Optical Amplifiers*, March 1996.

[NFOEC-1] Technical Paper, *"IP over WDM" the Missing Link*, P. Bonenfant, A. Rodrigues-Moral, J. Manchester, Lucent Technologies, A. McGuire, BT Laboratories, September 1999.

[OIF-1] OIF, The Optical Internetworking Forum Kick-Off Meeting Presentation, May 1998.

[OIF-2] OIF Contribution, *A Proposal to Use POS as Physical Layer up to OC-192c, OIF99.002.2*, January 1999.

[OIF-3] OIF Contribution, *IP Centric Control and Signaling for Optical Lightpaths*, January 2000.

# 3

# Optical Networking Technology Fundamentals

**U**nderstanding how to deliver Internet Protocol (IP) and IP services over an optical network infrastructure requires an in-depth understanding of the wide range of technologies used for building the appropriate optical network.

In this chapter, we provide the necessary technology background for understanding how next-generation networks are designed and deployed. We will describe optical transmission technologies and issues, as well as list the most common optical transmission systems and their capabilities. We will then cover the most commonly used data transmission technologies and will point out general concepts for building reliable networks.

## Optical Transmission Technologies

The Optical Transport Network (OTN) provides the basis for the service infrastructure delivering commonly IP services. The OTN facilitates a variety of functions.

The transmission capacity over a single fiber is increased by using Wavelength Division Multiplexing (WDM) technology to transmit multiple wavelengths over a single fiber and to achieve transmission capacities on the order of terabits per second.

The use of optical amplifiers makes it possible to transmit optical signals up to several hundreds of kilometers without any electro/optical conversion. Amplifier sites that are part of a DWDM system may also incorporate add/drop functions to allow nodes connected to this site to terminate some of the wavelengths.

Because the OTN commonly consists of a mesh of point-to-point (p-t-p) DWDM connections, there is a need for intelligent nodes at the DWDM junctions. That is where wavelength routers come into play. They switch wavelengths dynamically and provision end-to-end wavelengths throughout the network.

In addition to all of these architectural issues, influences to optical signal transmission must be carefully considered to ensure proper operation.

## Wavelength Division Multiplexing (WDM)

WDM combines multiple signals, each at a different carrier wavelength to increase fiber capacity. The first WDM networks deployed two wavelengths: one in the 1310-nm window and the other in the 1550-nm window. Today's WDM systems utilize 16, 32, 128, or more wavelengths in the 1550-nm window and are commonly called ***Dense Wavelength Division Multiplexing*** (DWDM) networks because of the densely packed wavelengths.

The International Telecommunication Union (ITU) has been active in trying to standardize a set of wavelengths for the use of WDM networks. This is necessary to ensure interoperability between systems from different vendors. The ITU-T recommendation G.692, "Optical Interfaces for Multi-Channel Systems with Optical Amplifiers," defines the optical parameters for DWDM systems used in interoffice and long-haul applications. G.692 specifies optical line systems with a maximum number of channels of 4, 8, 16, 32, or more wavelength transporting STM-4, STM-16, or STM-64 signals over a single fiber, using either uni- or bidirectional optical transmission.

The range of standardized channel grids includes 50-, 100-, 200-, and 1000-GHz spacing, providing DWDM system developers with enough flexibility to optimize the system specific to their needs. The DWDM band is suggested to be placed in the 1550-nm window with a nominal center frequency of 193.10 THz (1544.53 nm). The 1550-nm window is preferred over the 1310-nm window because of the lower fiber attenuation and the availability of erbium-doped fiber amplifiers (EDFAs), facilitating a pass-band covering the entire 1550-nm window. Figure 3–1 shows how wavelengths are placed into the 1550-nm win-

dow using the standardized 100-GHz spacing. By using this wavelength allocation scheme, a maximum of 41 channels can be transmitted across a single fiber.

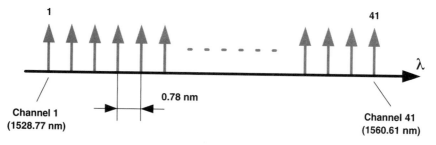

**Figure 3–1**    100-GHz spacing specified in ITU-T G.692

The recommendation suggests various wavelength allocations, starting at
1528.77 nm and ending at 1560.61 nm. The 1560.61-nm boundary will be
moved as technology evolves.

In theory, the signal rate for each wavelength and the deployed number of
channels can be increased as desired. An increased signal rate leads to a broader
channel spectrum, and an increased number of channels leads to a smaller channel spacing. Wavelength channels, therefore, move together, and the practical
limit is defined by the applied laser and filter technology, so that the channel
spectrums do not overlap. This issue is illustrated in Figure 3–2.

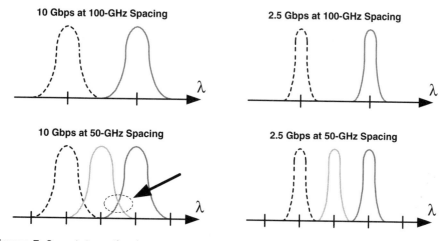

**Figure 3–2**    Interaction between channel spacing and signal line rate

In addition, WDM networks and their components have to deal with all the influences to optical signal transmission, including impairments such as attenuation, dispersion, fiber nonlinearities, and optical signal-to-noise ratio (OSNR).

The reasons and impacts of these impairments are described in the following section. For further details see *Optical Networks—A Practical Perspective*, [RAM-1].

## Attenuation

When transmitting an optical signal across a fiber, the optical power level gets attenuated. There are several loss mechanisms. The most important one is ***Rayleigh scattering***. Rayleigh scattering is caused by fluctuations in the density of the silicon used to produce the fiber. The introduced loss is decreasing as the wavelength of the optical signal increases. As a result, a loss at about 0.2 dB/km can be achieved in the commonly used 1550-nm window.

Figure 3–3 illustrates the fiber attenuation as a function of the operating wavelength. At least two "optical windows" can be defined. The first is a 1310-nm window, commonly used for optical interfaces of networking equipment. The second is a 1550-nm window, which is the preferred part of the wavelength spectrum for DWDM systems. In the remaining parts of the wavelength spectrum, the attenuation is very high and the achievable transmission distances, therefore, very small. On the other hand, in the 850-nm region, LEDs can be used instead of lasers, making it possible to produce cost-effective, optical short-haul interfaces.

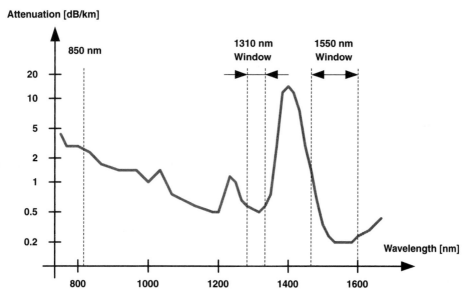

**Figure 3–3**  Typical fiber attenuation, depending on the operating wavelength

## Dispersion

***Dispersion*** is the overall name for all effects forcing different colors of an optical signal traveling at different speeds through the fiber. As a result, the signal gets smeared, as is shown in Figure 3–4. Very high dispersion can lead to overlapping pulses, making it impossible to extract the information at the receiver.

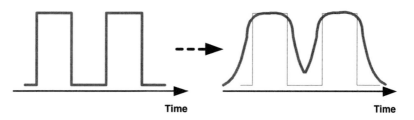

**Figure 3–4**  Dispersion forces the optical signal to get smeared, causing problems at the receiver

In general, there are three types of dispersion. The first one, ***modal dispersion*** (MD), is caused by the effect that different modes traveling in a multimode fiber (MMF) have different speeds. It is the effect limiting the achievable transmission distances with MMFs but does not influence DWDM network design, where single-mode fibers (SMFs) are used.

The second type of dispersion is ***polarization-mode dispersion*** (PMD), introduced by fibers having a core with imperfect concentricity. As a consequence, different polarizations of the optical signal have different propagation delays.

The third and most important type is ***chromatic dispersion*** (CD), which is caused by the frequency-dependent refractive index of the silicon used to produce the fiber and the frequency-dependent power distribution between the core and the cladding. Both components make different parts of a channel's wavelength spectrum travel at different speeds.

Figure 3–5 shows the chromatic dispersion as a function of the wavelength for both a standard SMF, specified in ITU-T G.652 [ITU-6], and a dispersion-shifted fiber (DSF), specified in ITU-T G.653 [ITU-7]. Dispersion is typically measured in ps/(nm*km), where ***ps*** refers to the time spread of the pulse, ***nm*** refers to the spectral width of the optical source used in the transmitter, and ***km*** denotes the length of the fiber connection.

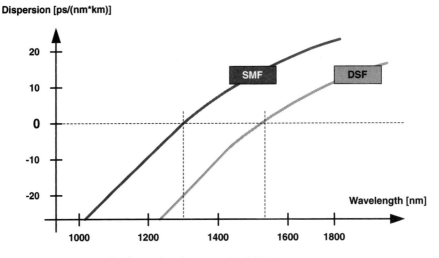

**Figure 3–5**    Chromatic dispersion in a standard SMF

Dispersion is the limiting factor for long distances. Attenuation can easily be handled through optical amplifiers but dispersion compensation requires more complex mechanisms. One solution to overcome the problem of dispersion is the use of special fiber types. For example, a DSF has the wavelength with a dispersion nearly equal to zero moved into the 1550-nm window; thus, influence of dispersion to the DWDM signal is very limited.

When reducing chromatic dispersion, it must be especially kept in mind that too little chromatic dispersion increases the influence of nonlinear effects, such as four-wave mixing (FWM), as defined in this chapter. Therefore, it is better to use standard SMFs and perform dispersion compensation in the several sites of the DWDM system.

## Nonlinear Effects

As is often the case in electronics, we can assume linear behavior for a system as long as we use low voltages, currents, or low optical power levels when talking about optical transmission over a fiber. So, in the case of having low optical power inserted by the transmitter, the refractive index and attenuation of a fiber is independent of the optical signal power.

Most DWDM systems use high optical power levels; therefore, nonlinear effects have to be considered. In general, there are two groups of nonlinear effects. The first one is caused through scattering effects in the fiber, and the second one is introduced because of the optical power-dependent refractive index.

Scattering effects lead to optical power being transferred from one wavelength to another. The wavelength loosing power acts as a pump wavelength and amplifies another wavelength. When having a single wavelength signal across a fiber, *stimulated Brillouin scattering* (SBS) forces optical power to be transferred onto a signal traveling in the opposite direction. This signal flowing toward the transmitter must then be removed through an isolator. Considering a DWDM signal, *stimulated Raman scattering* (SRS) evokes lower wavelengths amplifying higher wavelengths. The result is a DWDM spectrum with some highly powered channels and some low-powered channels. The channels with low power may not be received properly because their OSNR is very low.

FWM caused by the power-dependent refractive index creates new wavelengths that are the mixing products of the DWDM channel wavelengths. The new wavelengths interfere with the original wavelengths, and as a result, optical power is transferred between several wavelengths. This also provokes OSNR problems because of low-powered channels.

*Self-phase modulation* (SPM) and *cross-phase modulation* (CPM) are also caused by the power-dependent refractive index. These two effects do not transfer optical power. Instead, they affect the phase and optical spectrum of the optical signal components and lead to a higher dispersion penalty.

To reduce the influence of nonlinear effect, the optical power per channel should be kept under the SBS and SRS threshold, and an unequal channel spacing can be applied to avoid creation of new wavelengths through FWM.

## Optical Signal-to-Noise Ratio (OSNR)

The OSNR denotes the power level difference between the channel signal and the optical noise. At each optical amplifier, noise called *amplified spontaneous emission* (ASE) is added to the original signal. This noise is accumulated at each amplifier stage. This may result in the ratio between the original signal and the added noise becoming too low, so that the receiver cannot extract the transmitted signal.

Especially in 10-Gbps DWDM systems, OSNR must be taken care of. Some DWDM systems support measuring the ONSR at several points in the network (for example, at the transmitter), at intermediate amplifier sites, and at the receiver in order to monitor the OSNR in the network continuously.

# Optical Transmission Systems

## Optical Amplifiers

As an optical signal propagates through an optical fiber, it gets attenuated, and the optical power level decreases. An optical transmission system, therefore, must be designed properly to ensure that enough optical power arrives at the receiver so that the signal can be detected and extracted correctly. Traditionally, regenerators were used to deploy connections across distances above 40 km. Several regenerators were placed between the transmitter and receiver. The regenerator converted the optical signal with the low optical power level back to the electrical domain, reshaped and retimed it, and converted it back into the optical domain. The drawback of this solution is that there is a regenerator required per wavelength, and there is a different regenerator for each bit rate and modulation scheme.

As technology evolved, optical amplifiers have been developed that offer a lot of advantages. Optical amplifiers are bit rate and modulation scheme independent. As a consequence, it is easier to upgrade to higher bit rates or change to a client signal with different modulation. Moreover, optical amplifiers have a large-gain bandwidth, making it possible to amplify the entire DWDM band at once, which reduces the amount of required equipment for long-distance connections dramatically.

The theory behind the several types of optical amplifiers is described in depth in the book *Optical Networks—A Practical Perspective* [RAM-1]. The essential information has been extracted to provide an overview in the following sections.

### Erbium-Doped Fiber Amplifiers (EDFAs)

An EDFA is a purely optical device using a silicon fiber doped with ionized $Er^{3+}$ erbium atoms. The optical input signal is pumped using a pump laser typically operating at a wavelength of 980 nm or 1480 nm. A wavelength-selective coupler combines the input signal with the pump laser and introduces the resulting signal into the fiber. At the output, the pump signal is separated from the amplified signal. Typically, an isolator is used at the output of the EDFA to prevent reflections into the amplifier, which can make the amplifier operate as a laser. Figure 3–6 shows the structure of an EDFA.

By doping the fiber with erbium ions, high atom concentration at high-energy levels is achieved. This effect is called *population inversion*. Through introducing the pump signal, atoms transit between the different energy bands. The atoms moving from high-energy bands down to low-energy bands are emitting photons, which then amplify the optical signal traveling through the amplifier. This effect is called *stimulated emission*.

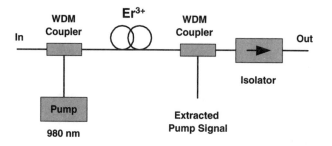

**Figure 3–6**    Structure of an erbium-doped fiber amplifier

Standard EDFAs use a pump laser operating at 980 nm. A high population inversion that delivers high gain can be achieved. Furthermore, the amplified spontaneous emission (ASE), which adds noise to the amplified signal, is very low. The drawback of these EDFAs is the fact that the gain curve is not perfectly flat. A typical EDFA gain curve can be seen in Figure 3–7, which is taken from the whitepaper "Today's Optical Amplifier—The Cornerstone of Tomorrow's Optical Layer" [ALC-1]. The wavelengths below the 1540-nm region especially are diversely amplified, which can lead to a condition where some wavelengths of an DWDM signal do not have a strong enough ONSR and may not be correctly received at the end of the transmission.

To overcome this problem, the DWDM band is typically reduced to the upper half of the EDFA's passband to exclude the 1530-nm to 1542-nm region. In addition, each DWDM channel is selectively powered at the transmitter to counterbalance the EDFA's not perfectly flat gain curve and to ensure a flat DWDM spectrum where all channels have the same OSNR at the receiving side.

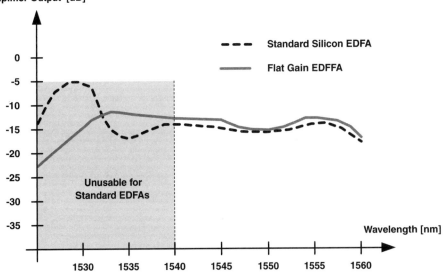

**Figure 3–7**    EDFFAs provide a much flatter gain curve than standard silicon EDFAs

In addition, flat-gain EDFAs have been developed that use a fluoride glass fiber instead of a silicon fiber. These amplifiers are using a 1480-nm pump laser and are called ***erbium-doped fluoride fiber amplifiers*** (EDFFAs). These amplifiers are controlling the gain curve internally to ensure it stays very flat. Therefore, a DWDM signal may use the whole EDFA passband from 1530 nm to 1560 nm. Figure 3–7 compares the gain curve of standard silicon EDFAs and flat-gain EDFFAs.

Another advantage of the flatter gain curve is that power balancing for each channel of the DWDM system is not as complicated as it is when using standard silicon EDFAs. The drawback of the EDFFAs is the higher noise figure caused by using a 1480-nm pump laser. This laser achieves lower population inversion and, therefore, a higher ASE.

In practice, optical amplifiers are commonly implemented using a two-stage design. Figure 3–8 shows such an amplifier using a standard silicon EDFA for the first stage because of its low noise figure and a flat-gain EDFA for the second stage to ensure an overall flat gain of the amplifier.

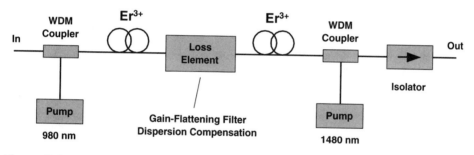

**Figure 3–8**    Structure of a two-stage EDFA

EDFAs are commonly used in the industry. They offer a wide amplification bandwidth. The noise figure is very low and typically only about 2 dB higher than the theoretical quantum limit of 3 dB. Furthermore, EDFAs are easy to use because of their simplicity and compactness.

EDFAs implemented in today's networks have a gain of about 25 dB or 30 dB and a gain bandwidth of about 35 nm. They facilitate spans up to 800 km without the use of electronic regenerators.

## Praseodynamium-Doped Fiber Amplifiers (PDFAs)

PDFAs use a similar concept but are optimized to operate in the 1310-nm window. They use a 1017-nm pump laser and fluoride fiber instead of silicon fiber. PDFAs offer the possibility to deploy long-haul single-channel applications using optical amplifiers and cheap 1310-nm interfaces as the networking equipment. PDFAs may also be used to expand the capacity of a DWDM system through the use of the 1310-nm band in addition to the 1550-nm band.

### Amplifier Control

When designing a DWDM system, a couple of issues have to be taken into account regarding the use of optical amplifiers. Typically, high power levels are to be achieved because the OSNR of the signal remains high. However, excessively high optical power at the input of an optical amplifier makes the amplifier go into the saturation mode, and the gain drops.

Furthermore, when using EDFAs, the gain curve is not perfectly flat. Therefore, certain channels of the DWDM signal are to be powered differently at the output amplifier stage of the transmitting side.

To cope with these issues, an automatic gain control (AGC) should be used. An AGC makes it possible to hold the output power of each channel constant, regardless of its input power. The AGC is monitoring the optical output signal and adjusts the pump laser as it is required, using an optical feedback loop.

Figure 3–9 shows how an AGC may be used in a p-t-p DWDM system. An optical supervisory channel (OSC) is used to exchange information about the optical signal power levels and other control information between the several geographically distributed sites. The OSC is extracted from the DWDM signal and processed by the AGC logic. The OSC uses a fixed wavelength outside the DWDM band. As the ITU-T recommendation G.692 specifies, the 1550-nm window to be used for transmitting the DWDM signal, typically a wavelength in the 1310-nm or 1600-nm region, is used for the OSC.

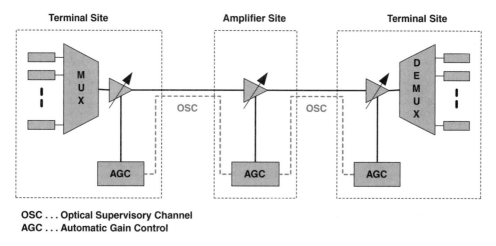

OSC . . . Optical Supervisory Channel
AGC . . . Automatic Gain Control

**Figure 3–9**    The automatic gain control ensures constant output power

## DWDM Systems

DWDM systems are used to transmit multiple wavelengths over a single fiber. These multiple channels acting as "virtual fibers" can be used to transmit any type of signal, such as Synchronous Digital Hierarchy (SDH), Asynchronous Transfer Mode (ATM), or IP traffic. These traffic sources are connected via standard optical interfaces.

There are certain distinctions between p-t-p and ring DWDM systems. Both can either be unidirectional or bidirectional systems. Unidirectional DWDM systems use all available wavelength channels of a fiber connection for either transmitting or receiving data. Bidirectional systems have some receiving and some transmitting wavelengths on the same fiber.

P-t-p systems are mostly used for increasing fiber capacity of long-haul connections. P-t-p systems, as shown in Figure 3–10, have two ***multiwavelength trunk interfaces*** for deploying a dual p-t-p fiber interconnection to provide link redundancy and protection capabilities. A number of single-wavelength interfaces are used to connect IP routers, ATM switches, or Synchronous Optical Network (SONET)/SDH equipment. The single-wavelength input ports have a ***modulator*** to transform the wavelength to the appropriate DWDM channel wavelength. The ***optical combiner*** multiplexes these wavelengths together, which are then sent over the multiwavelength trunk. At the receiving side, each chan-

nel is demultiplexed through an ***optical filter. Regenerators*** reshape and ream-
plify the filtered single-wavelength signal.

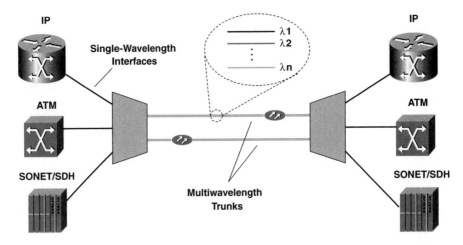

**Figure 3–10**    Point-to-point DWDM systems are used to increase fiber capacity

P-t-p systems typically use ***1 + 1 protection*** to handle fiber or wavelength
channel failures. The two DWDM terminals are interconnected via two fibers.
Each wavelength is sent over both fibers; the receiving terminal compares both
optical signals and uses the better one.

Figure 3–11 shows a bidirectional four-channel p-t-p DWDM system.
Channels 1 and 2 are used for transmitting data from the left to the right, and
channels 3 and 4 are used for transmitting data from the right to the left. There
is the same wavelength allocation on both fibers to provide 1 + 1 protection, as
mentioned before.

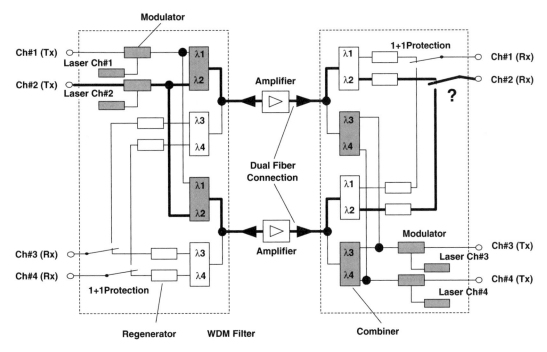

**Figure 3–11**    Block diagram of two connected four-channel DWDM p-t-p system terminals

Especially for Metropolitan Area Network (MAN) applications, ring systems with add/drop functionality are used. These systems have four multiwavelength trunk interfaces for deploying dual fiber rings with enhanced protection functionality. Numerous single-wavelength interfaces are acting as add/drop ports and are used either to connect to IP routers and ATM switches or to ring cross-connects.

As can be seen in Figure 3–12, DWDM rings can provide any virtual topology. P-t-p, ring, hub-and-spoke, or mesh networks can be deployed by manually configuring how the wavelengths are terminated.

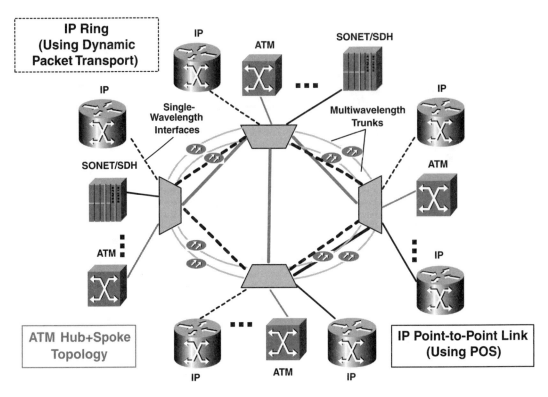

**Figure 3–12**    DWDM ring systems providing any virtual topology

Figure 3–13 shows a four-channel DWDM ring node. DWDM filters extract the wavelength channels out of the receiving multiwavelength trunks and feed them into the switching matrix. The switching matrix utilizes the configured add/drop functionality by forwarding an incoming wavelength either to a local drop port or to the optical combiner of the transmitting multiwavelength trunk. A channel at the combiner input can either be a pass-through channel coming from the optical filter or a signal from one of the local add ports.

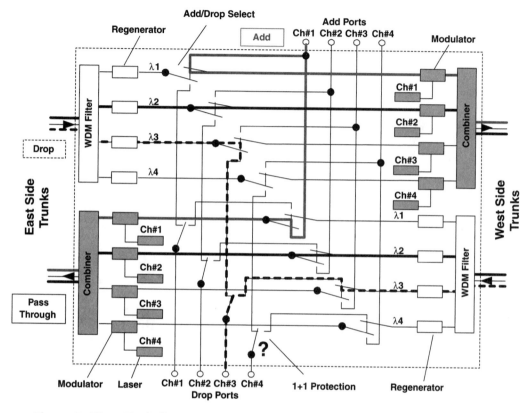

**Figure 3–13**    Block diagram of a four-channel DWDM ring systems terminal

Again, 1 + 1 protection is deployed by allocating a wavelength on both fibers and a comparison of the optical signals at the receiving node. By deploying DWDM systems, the fiber capacity can be easily increased. Networks become very scalable in terms of "raw bandwidth." In case more bandwidth is required, only one or more DWDM channels have to be added.

## Wavelength Routers

When implementing an intelligent optical network to overcome the static provisioning of legacy DWDM systems, wavelength routers are required. Wavelength routers are also often referred to as *optical cross-connects* (OXCs) or *wavelength cross-connects* (WXCs).

A wavelength router is a wavelength switching cross-connect providing optical interfaces. It has enhanced routing intelligence implemented in the network element management unit. The **network element management unit** controls and manages how the several wavelength channels are switched in its switching backplane. A wavelength router has **trunk ports** that connect to other wavelength routers or optical add/drop multiplexer (OADM) and **local ports**, used to connect to traffic sources or sinks, such as IP routers or ATM switches. End-to-end optical paths switched through the optical network are terminated at local ports.

Wavelength routers have several features. They are delivering end-to-end connections, which are then used for building the desired topology (mesh, ring, or p-t-p) of the IP network. They make it possible to provision end-to-end wavelength paths dynamically across the OTN. Enhanced survivability is delivered by the intelligent and individual wavelength routing functionality. The connection placement is reoptimized as the network utilization and traffic patterns change; thus, an optimum network capacity utilization is achieved.

The major building blocks of wavelength routers are the switching backplane, the I/O system, and the wavelength routing engine.

Concerning the **switching backplane**, three types of wavelength routers can be distinguished. The first-generation wavelength routers are electrical wavelength routers, using electrical backplanes. Enhanced products, so-called hybrid wavelength routers, have the backplane split into an electrical and an optical backplane. The third category includes optical wavelength routers, using a fully optical backplane. Most implementations found today are based on electrical backplanes. Hybrid implementations can be seen as an evolutionary path to the all-optical backplanes. Products seen in the next few years will based be in most cases on hybrid wavelength router architectures. The all-optical wavelength router is available in simple products. It seems as if it will become the dominant solution but still has to go through further technological evolutions.

The **I/O system** provides access to the backplane via standard optical interfaces. It performs electro/optical transformation in case of electrical or hybrid wavelength routers. In optical wavelength routers, the I/O system is responsible for wavelength transformation. There are two types of standard optical interfaces. Single-wavelength interfaces are used as local ports to connect routers, ATM switches, or SDH equipment. For deploying trunk connections, multiwavelength interfaces are provided through WDM multiplexers and demulti-plexers.

The ***wavelength routing engine*** is implemented in the network element man-agement unit of the switching matrix. Routing is based on link state routing protocols such as OSPF and Intermediate System to Intermediate System (IS-IS) protocol. At this time, there is no standard wavelength routing protocol pro-posed by any standardization body. The criteria for implementing a wavelength routing protocol algorithm is that every wavelength router must know the entire topology of the network. This is the reason why ***link-state protocol*** must be used. In addition to link-state messages, resource information has to be exchanged to make rerouting possible.

The wavelength routing engine performs two types of signaling. The first one is used for exchanging information between routers and running the routing protocol. The second type of signaling is used between the switching backplane and the routing engine to manage the local wavelength-switching process. The wavelength routing engine also provides the interface to the management tool and calculates alternate paths for mesh protection.

### Electrical Wavelength Routers

Electrical wavelength routers terminate the wavelengths at E/O converters, transform them into electrical signals, and switch them in the electrical domain. The backplane of an electrical wavelength router consists of an electrical non-blocking crossbar-switching matrix. Today's implementations have OC-48/STM-16 granularity, thus providing 2.5-Gbps wavelength services. Figure 3–14 shows a block diagram of an electrical wavelength router. Ingress traffic at any port can be switched through the matrix to any egress port.

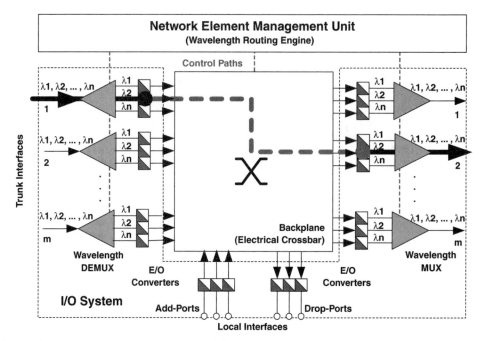

**Figure 3–14**    Block diagram of an electrical wavelength router

The crossbar matrix establishes the connection between the desired port pair. A framing module at each incoming and outgoing port is responsible for removing/adding the frame structure of the standard optical interface and for providing fixed-length packets to the matrix queues.

## Hybrid Wavelength Routers

To optimize the transport of transiting wavelength paths, the switching matrix of hybrid wavelength routers is split into two parts, as shown in Figure 3–15. Some of the trunks are connected to the optical switching matrix, which is used for transparently switching a wavelength from one trunk to the other. Only in cases where wavelength transformation is required—because the DWDM system at the desired output trunk has already allocated the wavelength used at the incoming trunk—is the electrical switching matrix utilized.

In addition, if the desired output trunk is not directly connected to the optical switching matrix or signal regeneration is necessary, the electrical switching matrix is utilized.

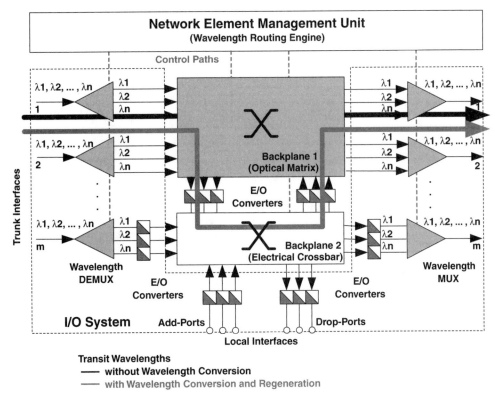

**Figure 3–15**    Block diagram of a hybrid wavelength router

## Optical Wavelength Routers

All optical networks deploy optical wavelength routers. No electro/optical conversion takes place in the wavelength router. The switching matrix is fully optical and nonblocking. The block diagram of an optical wavelength router is shown in Figure 3–16. The optical wavelength router switches wavelengths transparently and does not care about framing and signal regeneration, which makes designing the optical network much more difficult.

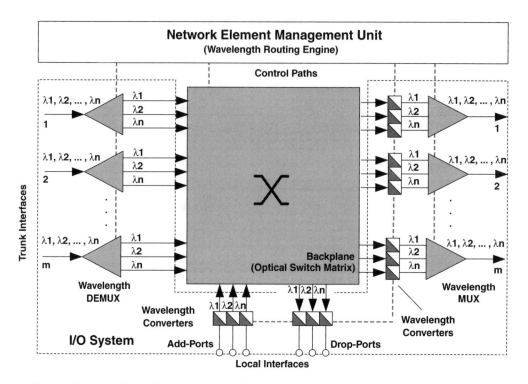

**Figure 3–16**    Block diagram of an optical wavelength router

*Wavelength conversion* plays a significant role in OXCs because, in the absence of wavelength conversion, a lightpath must be assigned the same wavelength on all links along its path through the network. With wavelength conversion, it may be assigned to different wavelengths on different links along its path. This is necessary for improving link and wavelength utilization in optical networks.

Figure 3–17 shows a wavelength allocation example. If the wavelength routers provide wavelength conversion, the optical paths from A to C and from F to C can use the link between B and C.

Without wavelength conversion, the path between F and C has to take the longer trace through node E and D if node F can assign only λ1 at the link to E.

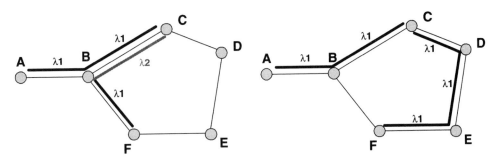

**Figure 3–17**    Wavelength allocation with or without wavelength conversion

An optical wavelength router with exactly two input and two output trunk ports, excluding any protection interfaces, acts as an OADM or wavelength add/drop multiplexer (WADM). Looking at Figure 3–18, you can see that there are only two trunk ports multiplexing and two trunk ports demultiplexing wavelengths. Instead of the other trunk ports, there are local add/drop ports having wavelength converters only for transforming the standard optical interface signals to the appropriate WDM channel wavelengths.

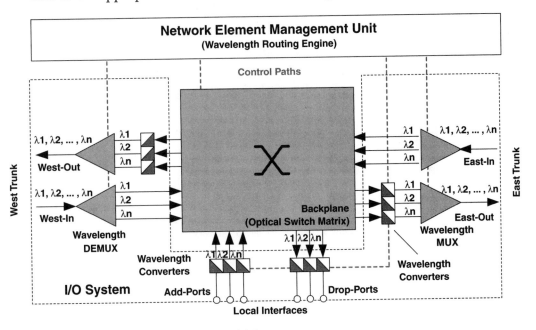

**Figure 3–18**    Optical add/drop multiplexer

Whereas electrical add/drop multiplexers must terminate the entire optical signal, OADMs customize traffic flow for maximum efficiency by terminating only a selected subset of the optical channels on the fiber. OADMs selectively drop or add specific wavelengths on a WDM fiber and provide them via local ports. Rather than demultiplexing all of the wavelengths through an optical coupler common in a long-haul system, only given wavelengths are added or dropped at a node. The other wavelengths are passed through the node optically. Furthermore, OADMs add and drop wavelengths irrespective of their content. Excepted from this are electrical ADMs, which monitor overhead bytes to select the part of the frame to be added or dropped.

# Data Transmission Technologies

The way to transport data, IP traffic for example, has fundamentally changed in optical networks. By removing the intermediate ATM and SONET/SDH layers, higher transport efficiency in terms of overhead is achieved, and more simple and streamlined network architectures can be created.

There are several encapsulation technologies used to transport IP data traffic across an optical network, such as Packet over SONET/SDH (POS), Dynamic Packet Transport (DPT), and Ethernet.

POS itself implies the use of SONET/SDH equipment. It is based only on standard SONET/SDH framing and defines encapsulation techniques used for p-t-p links.

As opposed to POS, resilient packet ring technologies such as DPT are defining encapsulation methods for ring topologies. In addition, protection mechanisms are included that may be compared with the protection switching functions from SONET/SDH rings.

Ethernet is no longer a Local Area Network (LAN) technology. It has already emerged to become a technology suitable for MAN and Wide Area Network (WAN) connections.

Ethernet is not going to be covered in the upcoming sections because the evolution from Ethernet LANs to Gigabit Ethernet or 10-Gigabit Ethernet WAN links is straightforward.

## Packet over SONET/SDH (POS)

POS allows the IP layer to be placed directly above the SONET/SDH layer, and, while offering Quality of Service (QoS) guarantees, it eliminates the overhead needed to run IP over ATM over SONET/SDH.

Because IP is a connectionless network layer protocol (while SONET/SDH is a physical layer protocol), the gap between layer 3 and layer 1 must be filled. This is achieved by the Point-to-Point Protocol (PPP), which is used for encapsulating IP packets into a data stream that is then mapped into SONET or SDH payloads. Therefore, POS requires a topology of *full-duplex p-t-p links*.

PPP consists of two protocol parts. The ***Link Control Protocol*** (LCP) establishes and tests the data link connection, while the ***Network Control Protocol*** (NCP) interfaces to the layer 3 protocol.

Figure 3–19 shows all necessary processes for mapping IP packets into SONET/SDH frames using PPP.

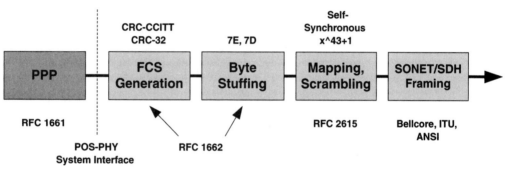

**Figure 3–19**   POS processes

POS implementations are based on the following Internet Engineering Task Force (IETF) documents:

- RFC 2615, "PPP over SONET/SDH" [IETF-1]
- RFC 1661, "The Point-to-Point Protocol (PPP)" [IETF-2]
- RFC 1662, "PPP in HDLC-like framing" [IETF-3]

The fundamentals of POS were standardized in RFC 1619. Because some issues of payload transparency and security appeared, the IETF "PPP Extensions" working group developed the IETF Draft "PPP over SONET/SDH" [IETF-4], which then became RFC 2615. RFC 2615 provides the basis for the following summary of POS. The paragraphs "Encapsulation" and "Framing" reflect RFC 1661 and RFC 1662, respectively.

### PPP Encapsulation

IP packets are mapped into SONET/SDH frames using PPP encapsulation. IP datagrams are filled into packets up to the size of the maximum transmission unit (MTU) and are encapsulated as shown in Figure 3–20.

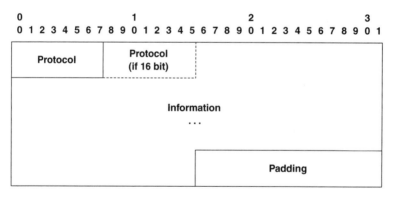

Maximum Transmission Unit (MTU) = Information + Padding = 1500 bytes

**Figure 3–20**    PPP encapsulation

The Protocol field is one or two bytes long, and its value identifies the datagram encapsulated in the Information field of the packet. Typical Protocol field values are shown in Table 3–1.

**Table 3–1**   PPP Protocol field values

| PROTOCOL FIELD | PROTOCOL NAME |
|---|---|
| 0x0001 | Padding Protocol |
| 0x0021 | Internet Protocol (IP) |
| 0xc021 | Link Control Protocol (LCP) |
| 0xc023 | Password Authentification Protocol (PAP) |
| 0xc025 | Link Quality Report |
| 0xc223 | Challenge Handshake Authent. Protocol (CHAP) |

The payload is carried in the ***Information*** field. It contains the datagram for the protocol specified in the Protocol field. The MTU defines the maximum length of the information field plus padding. The default MTU is specified within 1,500 bytes.

During the PPP encapsulation process, padding is appended to small datagrams received from the upper layer. In case of datagrams bigger than the MTU, the datagrams are fragmented and transmitted by using multiple PPP packets. At the receiving side, the upper layer protocol is responsible for distinguishing between padding and real data.

LCP packets are used to establish the communication over the p-t-p link. The LCP packets ensure that the link is configured and tested correctly.

NCP packets are used to choose and configure the upper layer protocol. After the successful configuration, upper layer datagrams can be transmitted over the link within PPP packets.

The link teardown is also controlled through the use of LCP and NCP packets. As long as no LCP or NCP packets request to close the link, datagrams are sent across the link. Such LCP or NCP packets can be triggered by the upper layer protocol or by external events, such as an inactivity timer expiration or network operator intervention.

## Framing

The Highspeed Data Link Control Protocol (HDLC)-like frame structure is shown in Figure 3–21. This figure does not include bits inserted for synchronization or any bits or bytes inserted for transparency.

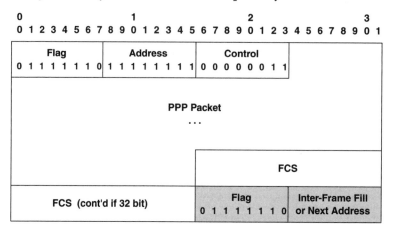

**Figure 3–21**    PPP HDLC-like frame structure

To delineate the frames, each frame has an 8-bit-long *Flag* sequence at the beginning and at the end. The Flag is defined as the binary sequence "01111110" (0x7E). Between two frames, only one Flag is required. Two consecutive Flags denote an empty frame, which is simply deleted and not counted as a frame check sequence (FCS) error.

HDLC defines an 8-bit-long *Address* field. Because POS is used only for p-t-p links, there is no addressing functionality required within HDLC. Therefore, the Address field is set to the "All-Stations Address," which is the binary sequence "11111111" (0xFF). Nevertheless, the Address field must be processed to always recognize the All-Stations Address.

The 8-bit-long *Control* field is used in HDLC to implement some control functions. Within POS, it contains the binary sequence "00000011" (0x03), which is called the *unnumbered information* (UI) command with the poll/final (P/F) bit set to zero.

The FCS is calculated over the whole frame, including the Address, Control, Protocol, Information, and Padding fields (but not the FCS field itself) and any bits or bytes inserted by the stuffing mechanism to ensure transparency, as will be discussed next. RFC 1662 defines a 16-bit-long FCS field at the end of the frame, but the length can also be set to 32 bits.

Bytes equal to the Flag in the information field would lead to wrong frame alignment. To avoid this, a **byte stuffing** procedure is used to ensure data transparency. The binary sequence "01111101" (0x7D) is used as Control Escape sequence. Byte stuffing is applied after FCS computation. The entire frame between two Flags is examined, and each Control Escape and Flag sequence is replaced by a 2-byte sequence consisting of the Control Escape byte, followed by the original byte exclusive-or'd (XOR'd) with "0x20." At the receiving side, before FCS computations, each Control Escape sequence is removed, and the subsequent byte is XOR'd with "0x20," unless it is equal to the Flag, which aborts a frame.

Examples:

```
0x7e is encoded as 0x7d, 0x5e    (Flag Sequence)
0x7d is encoded as 0x7d, 0x5d    (Control Escape)
```

The standard **LCP sync configuration** defaults apply to SONET/SDH links. Therefore, the following configuration options are recommended when implementing POS interfaces:

- Magic Number (part of PPP header to detect loop back)
- No Address and Control Field Compression
- No Protocol Field Compression
- 32-bit FCS

### Interface Format

HDLC-like framing is used to provide an octet interface to the physical layer (SONET/SDH). The **octet stream** is mapped into the SPE/AUG within the SPE/AUG octet boundaries. Binary line coding is used for SONET/SDH interfaces.

In this context, it is important to note that long all-zero sequences can cause problems within SONET/SDH networks. For example, SONET/SDH receivers often recover their clock from the incoming data stream to ensure proper synchronization throughout the network. In case of a long all-zero sequence, the phase-locked loop system of the receiver would come out of sync, which then would lead to incorrect timing for the SONET/SDH node and, furthermore, for all downstream SONET/SDH nodes.

Another important feature is protection against malicious user attacks trying to disrupt the SONET/SDH signal transmission. The solution is payload scrambling, which is applied to ensure constantly appearing transitions (zeros to ones and ones to zeros). The scrambling mode is indicated via the ***Path Signal Label*** represented by the C2 path overhead byte. To provide backward compatibility to POS implementations compliant to RFC 1916, the payload scrambler can be turned off by setting the C2 byte to "0xCF" (Table 3–2).

**Table 3–2**    Scrambling Mode Indication with C2

| C2 | SCRAMBLING |
|---|---|
| 0x16 | ON |
| 0xCF | OFF |

The multiframe indicator H4 of the path overhead is unused within the mapping process and is set to "0x00."

### Transmission Rate

RFC 2615 specifies only POS interfaces using concatenated SONET/SDH payloads (STS-Nc/STM-Nc). Concatenated payloads are created through linking multiple STS-1s/STM-1s together in fixed-phase alignment and use all STS-1s/STM-1s as one big container. The basic POS transmission rate is 155.52 Mbps (Table 3–3), which equals the OC-3c/STM-1 interface.

The available transmission capacity is derived as shown in Equation 3–1 or 3–2.

$$TC\,(STS-3c) = 9\,rows*(90*3-3*3-1)\,columns*\frac{1}{125us}*8\,bits\,/\,byte = 149,76\,Mbps$$
$$\text{(3–1)}$$

$$TC\,(STM-1) = 9\,rows*(270*1-9*1-1)\,columns*\frac{1}{125us}*8\,bits\,/\,byte = 149,76\,Mbps$$
$$\text{(3–2)}$$

Higher bit rate signals OC-12c/STM-4c and OC-48c/STM-16c POS interfaces are available. OC-192c/STM-64c interfaces are not defined in RFC 2615. The implementation of OC-192c/STM64c POS interfaces is covered in the "PPP over SONET/SDH at OC-192c/STM-64c" section later.

**Table 3–3**    POS Transmission Rates

| BIT RATE [MBPS] | CAPACITY [MBPS] | SONET | SDH |
|---|---|---|---|
| 155.52 | 149.760 | OC-3c | STM-1 |
| 622.08 | 600.768 | OC-12c | STM-4c |
| 2488.32 | 2404.800 | OC-48c | STM-16c |
| 9953.28 | 9620.928 | OC-192c | STM-64c |

A technology overview of SONET/SDH is included later in this chapter, describing frame structure, multiplexing hierarchy, overhead functions, etc. Also, a detailed overhead calculation for POS, together with a comparison of the three IP transport technologies ATM, POS, and DPT is included.

### Payload Scrambling

As defined in RFC 2615 and shown in Figure 3–22, POS uses a $x^{43}+1$ scrambler consisting of a 43-bit shift register and an exclusive-or (XOR) gate. This scrambler is self-synchronous, so no clock or frame pulse is used during scrambling.

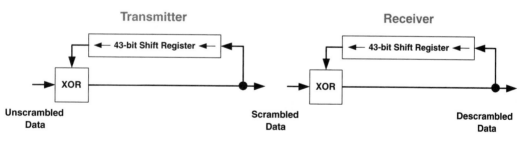

**Figure 3–22**    Payload scrambler

At the transmitting side, the unscrambled data stream is XOR'd with the output of the shift register. The output of the XOR gate delivering the scrambled data is fed back into the input of the shift register. At the receiving side, the scrambled data is XOR'd with the output of the shift register, and the output of the XOR gate delivering the descrambled data is fed back into the input of the shift register.

Unlike HDLC (where the HDLC FCS scrambling function is beginning with the least significant bit), the SONET/SDH scrambling function is beginning with the most significant bit. The scrambler runs continuously across the whole frame but excludes the POH bytes or any "fixed stuff" bytes of frame. Also, the scrambler runs continuously across multiple frames and is not reset per frame. The initial seed is randomly chosen by the transmitter to improve operational security. Consequently, the first 43 transmitted bits following startup or reframe operation will not be descrambled correctly.

## PPP over SONET/SDH at OC-192c/STM-64c

When implementing POS OC-192c/STM-64c interfaces, using the methods described in the paragraphs above, the processing of byte stuffing and destuffing becomes the challenging issue because of the high transmission rate of 10 Gbps. Especially, performing the comparison of each byte in the receiver to distinguish whether to remove the byte becomes very difficult to implement at such high line rates. There have been two approaches proposed to cope with this issue: a parallel implementation and word-oriented POS framing.

### PARALLEL IMPLEMENTATION

The first and most straightforward approach is to use a parallel implementation of the byte comparison. n bytes of the octet stream are compared in parallel, and therefore, processing can be performed at 1/8 n of the line rate. Thus, when using n = 16 for the 10-Gbps interface, a logic operating at a clock rate of only 77 MHz is required.

This quite low clock rate, together with a reasonable complexity and gate count, makes it possible to implement an OC-192c/STM-64c POS interface without requiring any changes to the above described POS specifications.

This method has been developed by the physical layer working group of the OIF for implementing a POS interface with a line rate of 10 Gbps. For implementation details, refer to [OIF-2].

### WORD-ORIENTED POS FRAMING

The second approach has been described in IETF draft, "PPP over SONET (SDH) at Rates from STS-1 (AU-3) to STS-192c (AU-4-64c/STM-64)" [IETF-5]. This draft proposes a 32-bit word-oriented POS framing for OC-192c/STM-64c POS interfaces and is summarized below.

Figure 3–23 shows the changes made by this proposal to the process flow (defined by RFC 2615) necessary for implementing OC-192c/STM-64c POS interfaces. The shaded blocks are defined by [IETF-5], and the remaining blocks have been left unchanged.

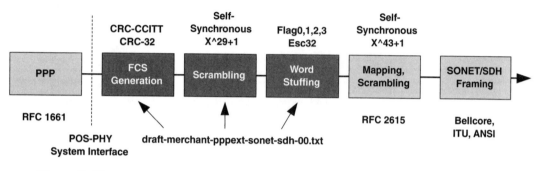

**Figure 3–23**    POS processes used for OC-192c/STM-64c interfaces

The proposed HDLC-like framing (HDLC-32), shown in Figure 3–24, uses a 32-bit-long word to encapsulate the PPP packets. It is responsible for packet delineation, data packet transparency, and error detection similar to the HDLC-like framing described in RFC 2615.

In addition, an adaptation function is implemented to convert the byte-oriented PPP packet to the word-oriented structure for HDLC-32.

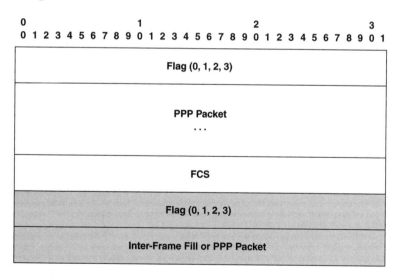

**Figure 3–24**    32-bit word-oriented HDLC-like frame structure (HDLC-32)

A 32-bit flag sequence is put at the beginning and end of each HDLC-32 frame. The FCS-32 is calculated over the whole PPP packet and is appended as a 32-bit FCS field at the end of the frame.

The byte stream is adapted to a 32-bit word stream by combining groups of four bytes to a word. If the byte stream is not a multiple of 4 bytes, the last word is padded with one, two, or three all-zero bytes. The padding is included in the FCS-32 calculation.

The number of pad bytes used for byte-to-word adaptation (0, 1, 2, or 3) is encoded in the choice of the corresponding Flag sequence (Flag0, Flag1, Flag2, or Flag3, respectively) that immediately follows the FCS-32 word for that packet.

The following Flag sequence values are proposed:

```
Flag0=0xE781CA34
Flag1=0xE781CA35
Flag2=0xE781CA36
Flag3=0xE781CA37
```

A self-synchronous $x^{29}+1$ payload scrambler should be applied after the FCS insertion to provide protection from possible malicious denial of service attacks by the sending of repeated sequences of flag or escape codes within the packet data, which could result in halving the effective bandwidth of the link. This scrambler operates independently of and in addition to the $x^{43}+1$ scrambler, which has been specified in RFC 2615 and is applied before mapping the word stream into the SONET/SDH payload.

*Word stuffing* is used to ensure correct frame alignment when data words equal the Flag sequence or the Escape sequence used for word stuffing. The Escape sequence used is:

```
Esc32=0xEB8DC638
```

After FCS-32 calculation and HDLC-32 payload scrambling, each Flag (Flag0, 1, 2 or 3) and Esc32 sequence shall be replaced by a two-word sequence consisting of the Esc32 sequence and the 32-bit word to be escaped XOR'd with the hexadecimal sequence "0x20202020."

In the receive direction, any occurrence of the Esc32 sequence shall be removed and the following word XOR'd with the hexadecimal sequence "0x20202020."

The only exception is the abort sequence "Esc32, Flag0," which indicates an invalid packet and should abort the processing of the receiving frame.

The scrambled word stream is then mapped into the SONET/SDH frame within its 32-bit boundaries. A different path signal label (C2) of the value "0x17" should be used to differentiate the HDLC-32 from the HDLC framing.

Because this method requires changes to RFC 2615, the parallel implementation approach might be the preferred one.

## Automatic Protection Switching (APS)

APS is the signaling protocol used in SONET rings for automatic circuit reconfiguration. By using the K1/K2 overhead bytes of the SONET frame nodes detecting a signal failure, other nodes are forced to switch over to the protection circuit. With APS the SONET recovery time limit of 60 ms could be achieved.

For an SDH ring, the ITU-T recommendation G.707 [ITU-3] defines Multiplex Section Protection (MSP) as a counterpart to SONET APS. Just as SONET and SDH are similar technologies, APS and MSP are also similar in their functions. Therefore, when using the term *APS* in the following discussion, APS and MSP are both implied.

OADMs used in optical networks may also be capable of processing the K1/K2 overhead bytes and, therefore, may also support APS. Thus, APS can be applied in optical rings directly interfacing to routers for deploying Unidirectional Path Switched Ring (UPSR) and BLSR functionality. The routers process the K1/K2 bytes and switch over to the protection interface in case of a signal failure, signaled through APS.

### APS CHANNEL PROTOCOL

The Bellcore SONET specification [BELL-1] defines the APS channel protocol, which is a bit-oriented protocol for exchanging and acknowledging protection-switching actions. There are two types of APS signaling defined:

1. Linear APS; used by routers acting as SONET line terminal equipment (LTE)
2. Ring APS; used by ADMs or OADMs in the ring

APS signaling is carried in the K1/K2 bytes of the protecting circuit. Figure 3–25 shows the bit usage of the K1/K2 bytes.

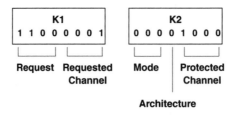

**Figure 3–25**    APS bit usage of K1/K2 overhead bytes

The first four bits of K1 define the type of request. APS action requests are grouped into three categories:

**1.** Automatically initiated (SF, SD)

**2.** Externally initiated (FS, MS, Lockout)

**3.** State request (WTR, DNR)

The remaining four bytes of K1 are used to indicate the channel (channel 1–14) for which the request is being made. This is especially needed in 1:N protection architectures, whereas N channels share one protection circuit.

The first three bits of K2 indicate the bidirectional mode with the binary sequence "101" or the unidirectional mode with the binary sequence "100." Bit 5 set to "0" represents the 1 + 1 architecture, and if set to "1," the 1:n architecture. Bits 4 to 1 represent the channel, which is switched onto the protection circuit (channel 1–14).

Table 3–4 summarizes the K1/K2 bit usage.

**Table 3–4**   K1/K2 Bit Usage

| K1 | | K2 | |
|---|---|---|---|
| **BIT 8..5** | **REQUEST** | **BIT 8..6** | **MODE** |
| 0xF | Lockout of protection | 101b | Bidirectional |
| 0xE | Forced switch (FS) | 100b | Unidirectional |
| 0xC | Signal failure (SF) | Bit 5 | Architecture |
| 0xA | Signal degrade (SD) | 0b | 1+1 |
| 0x8 | Manual switch (MS) | 1b | 1:n |
| 0x6 | Wait to restore (WTR) | **BITS 4..1** | **CHANNEL** |
| 0x2 | Reverse request (ACK) | 1..14d | Protected channel |
| 0x1 | Do not revert | | |
| 0x0 | No request | | |
| **BITS 4..1** | **CHANNEL** | | |
| 1..14d | Requested channel | | |

### CISCO'S APS IMPLEMENTATION

Cisco has defined a proprietary *APS Protect Group Protocol* used for communication between the process controlling the working and protection channels. This protocol makes it possible to have the working and protection interface on two separate routers, as shown in Figure 3–26.

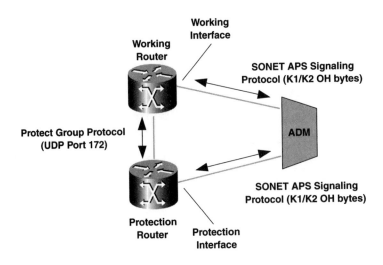

**Figure 3–26**    Cisco APS Protect Group Protocol between two routers

Both the working and the protection interface are controlled by a process called the *working* or *protection process*, respectively. Both processes communicate with each other by using the Protect Group Protocol. Protection switching may occur for the following reasons and is controlled by the protection process:

- Router/Interface Crash
- Signal Degradation (SD)
- Loss of Signal (SF)
- Manual Switch (MS)

In case of a failure, receive and transmit circuits are always switched in pair, initiated and controlled by the protection process.

The Protect Group Protocol defines a *protection group* for each protection relationship. This might be a pair of one working and one protection interface, but can also be a group of working interfaces and a single protection interface if 1:N protection is used. The UDP port 172 is used for exchanging the control messages between the protection process and working processes.

*Hellos* are sent by the protection process to each working process in the protect group to detect a failure. The Hello interval defaults to 1 second and the hold time to 3 seconds. In case of a failure, a `Linestate_Change` message is sent by the working process to the protection process. Then a `Working_Disable` message is sent by the protection process to disable the failed working interface, and the traffic is switched onto the protection interface. If the revertive mode is used and the failed working interface comes up again, a `Working_Enable` message is sent by the protection process to reenable the interface and switch back to the former failed working interface. *Acknowledgements* are sent by both processes to acknowledge messages.

If the communication between the two processes is lost, the protection process relies on the K1/K2 bytes received from the ADM.

For further information on Cisco's POS and APS implementations, refer to the whitepaper, "Cisco's Packet over SONET/SDH (POS) Technology Support" [CSCO-1].

### APS SWITCHOVER PROCESS

In case of a working fiber cut, the working interface will detect a loss of signal (SLOS alarm). The protection process signals a Signal Fail request and enables a channel on the protection interfaces between the ADM and the router. After this, the channel on the working interfaces is torn down. After this channel switchover, the working interface of the router connected to the ADM transits into idle state, and the protection interface takes over service.

### APS NETWORK LAYER INTERACTION

It is important to keep in mind that APS is a layer-1 signaling protocol and is handling only the interface switchover. This does not include rerouting at the network layer. Another fact is that the protection interface is in idle state during normal operation, which results in one heavily loaded interface (working) and one nonutilized interface (protection).

## Dynamic Packet Transport (DPT)

DPT is a transport technology developed by Cisco Systems that introduces a new layer-2 media access control (MAC) protocol called the *Spatial Reuse Protocol* (SRP). SRP enables the deployment of scalable and survivable optical IP packet rings.

As opposed to POS, which is a p-t-p technology, DPT makes it possible to build ring networks wherein data transmission across the ring does not stress intermediate nodes on the ring. Thus, DPT specifically addresses not only the challenges of metropolitan ring networks but also those of high-capacity backbones.

SRP has been proposed to the IETF in the informational RFC 2892 "The Cisco SRP MAC Layer Protocol" [IETF-8]. The following description of SRP is based on this RFC and the conference presentation [CSCO-4]. SRP is a layer-2 MAC protocol for LANs, MANs, and WANs.

DPT interfaces can easily be used to connect a router directly to SONET/ SDH equipment, DWDM systems, or dark fiber because SRP provides a standard SONET/SDH interface in its first implementation. This first implementation of DPT provides a great deal of flexibility because DPT rings can be deployed over one of these three physical layer choices. Moreover, a mix of these three "hybrid DPT rings" is possible. It is important to keep in mind that SRP is completely independent from the underlying physical layer and can be applied over any layer-1 technology.

DPT rings are *dual, counter-rotating fiber rings,* as shown in Figure 3–27. Both fibers are used for both transmitting data packets and control packets. There are several types of control packets, such as topology discovery packets, protection switching packets, and bandwidth usage control packets. As can be seen in Figure 3–27, the control packets of a specific ring are always sent on the other ring.

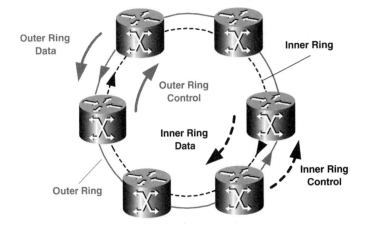

**Figure 3–27**   DPT uses dual, counter-rotating fiber rings, whereas both rings are used for data transmission

As opposed to well-known ring technologies, such as Token Ring or Fiber Distributed Data Interface (FDDI), SRP uses a destination-stripping mechanism. No token is used for admission control. The data is sent only between source and destination, which makes it possible to have concurrent traffic exchange on other parts of the ring. This so-called *spatial reuse capability* leads to an efficient bandwidth utilization and enables the total ring bandwidth to increase.

Figure 3–28 shows an example of an OC-48c/STM-16c DPT ring. Router A exchanges 1.5 Gbps of data with router D. At the same time, routers B and C can transmit data between each other up to 1 Gbps. Moreover, routers E and F can use the whole 2.5 Gbps of bandwidth on the left bottom part of the DPT. The total amount of data exchanged in this ring is 5 Gbps.

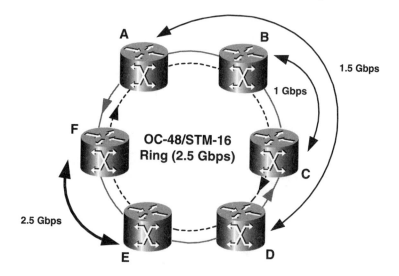

**Figure 3–28**  Spatial reuse capability

The *topology discovery* functionality of SRP facilitates "plug-and-play" operation and eliminates extensive initialization and preconfiguration for new nodes, as it is known with SONET/SDH rings.

## Encapsulation

In SRP, the IP datagrams are encapsulated in packets basically consisting of a header, payload, and FCS part. The frame structure is very similar to the Ethernet frame. All SRP packets have at least the first 16 bits of the header, known as the *SRP header*, in common. The SRP header is shown in Figure 3–29. This and all subsequent figures showing packet frame structure do not include layer-1 frame delineation because this depends on the given physical technology.

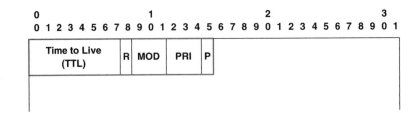

**Figure 3–29**    SRP header

The usage of the SRP header fields is summarized in Table 3–5.

**Table 3–5**    SRP Header Fields

| FIELD | BITS | VALUE | USAGE |
|-------|------|-------|-------|
| Time to Live (TTL) | 8 | | Hop Count |
| Ring Identifier (R) | 1 | 0 | Outer Ring |
| | | 1 | Inner Ring |
| Priority (PRI) | 3 | 0d..7d | Packet Priority |
| Mode | 3 | 000 | Reserved |
| | | 001 | Reserved |
| | | 010 | Reserved |
| | | 011 | ATM Cell |
| | | 100 | Control Message (Pass to Host) |
| | | 101 | Control Message (Locally Buffered for Host) |
| | | 110 | Usage Message |
| | | 111 | Packet Data |
| Parity (P) | 1 | | Odd Parity over MAC Header |

The maximum value for the DPT MTU is 9.216 bytes to be able to carry large IP MTUs. The minimum MTU is specified as 55 bytes because DPT also supports the transport of ATM cells over DPT; these DPT packets are 55 bytes long. The minimum MTU does not apply to the control packets because these are shorter.

### IP DATA PACKETS

Similar to Ethernet, a DPT data packet includes the source and destination MAC address required for performing address lookups, a protocol type field denoting the carried protocol, and an FCS for error detection. The MAC address used in DPT is a unique 48-bit address. The FCS calculation is performed over the whole packet, ignoring the 16-bit SRP header. The format of a data packet is shown in Figure 3–30.

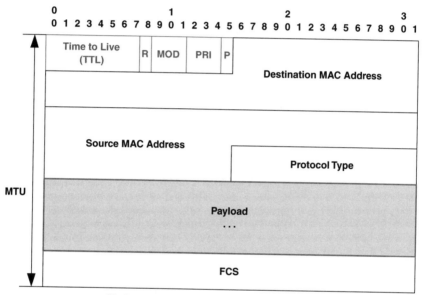

Maximum Transmission Unit (MTU) = 55 ... 9216 bytes

**Figure 3–30**    SRP data packet format used for IP packets

The specified values for the Protocol Type field are shown in Table 3–6.

**Table 3–6**    Possible Values of the Protocol Type Field

| FIELD | BITS | VALUE | USAGE |
|---|---|---|---|
| Protocol Type | 16 | 0x2007 | SRP Control |
|  |  | 0x0800 | Ipv4 |
|  |  | 0x0806 | ARP |

## ATM Data Packets

As can be seen in Table 3–5, there is a mode called **ATM Cell.** This mode represents the functionality to transport ATM cells across a DPT ring. This allows coexistence of IP routers and ATM switches on the same ring. The corresponding data packet format is shown in Figure 3–31.

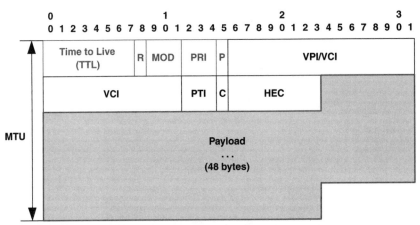

Maximum Transmission Unit (MTU) = 55 ... 9216 bytes

**Figure 3–31**    SRP data packet format used for ATM Cells

The ATM data packet does not include a FCS field. Data integrity is handled at the ATM adaptation layer (AAL).

## Control Packets

The SRP control packets have the format shown in Figure 3–32.

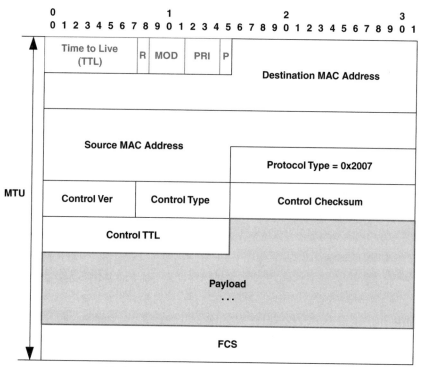

**Figure 3–32**     SRP control packet format

The value for the Protocol Type field is "0x2007," according to Table 3–6. The usage of the control packet MAC header fields is summarized in Table 3–7.

**Table 3–7**     SRP Control Packet Fields

| FIELD | BITS | VALUE | USAGE |
|-------|------|-------|-------|
| Control Ver | 8 | 0 | Control Type  Version |
| Control type | 8 | 0x01<br>0x02<br>0x03..0xFF | Topology Discovery<br>IPS Message<br>Reserved |
| Control TTL | 16 | | Control Layer  Hop-Count |

All control packets are sent p-t-p between adjacent nodes with the destination address "0x0" and the highest priority "0x7" and may be forwarded in a hop-by-hop manner, depending on the type of control packet.

#### Usage Packets

SRP usage packets are used to propagate allowed usage information. By periodically sending out usage packets, upstream nodes on the ring are informed about the current ring utilization. This information exchange is essential for the SRP fairness algorithm, described later. Usage packets also act as keepalives and therefore should be sent out every 106 µs. The usage packet format is shown in Figure 3–33.

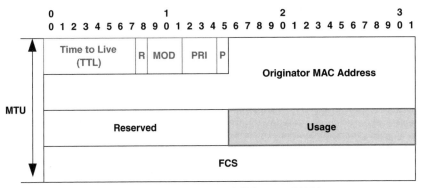

Maximum Transmission Unit (MTU) = max. 9216 bytes

**Figure 3–33**   SRP usage packet format

If a node does not receive usage packets for a certain keepalive timeout interval, it will trigger a keepalive timeout event, which then initiates protection-switching actions. Typically this timeout interval is set to 16 times the usage packet transmission interval.

## SRP Physical Layer Implementation

#### Framing

Although SRP is a physical-media-independent MAC layer protocol, the first implementation of SRP is based on SONET/SDH as the physical layer. This implementation uses the same flag delineation and octet stuffing mechanisms as already used in POS.

The binary sequence "01111110" (0x7E) is appended both at the beginning and end of each SRP packet to indicate the packet boundaries. The byte stuffing mechanism uses the binary sequence "01111101" (0x7D) as an escape character. This ensures that data bytes equal to the flag or escape character do not lead to wrong frame alignment.

### INTERFACE FORMAT

The data stream is mapped into the SPE/AUG of concatenated SONET/SDH frames. DPT interfaces can be used to connect to SONET/SDH ADMs, dark fiber, and DWDM terminals, thus providing a lot of flexibility for network designers.

### TRANSMISSION RATE

DPT is currently defined for OC-12c/STM-4c and OC-48c/STM-16c interfaces. To meet the full range of requirements of future optical networks in the backbone area, OC-192cSTM-64c interfaces will also be implemented.

## Packet Handling Procedures

### TOPOLOGY DISCOVERY

Each node performs topology discovery by sending out topology discovery packets on one or both rings. The packets are sent p-t-p to the neighbors, and each node appends its MAC binding information, consisting of the MAC address, the ring ID, the Wrap status, and updates the topology length field. A node updates its topology map of all stations and their Wrap status after receiving two identical topology packets that it has originated. That means that both discovery packets have traveled around the whole ring and have been processed by all nodes on the ring. The format of a topology discovery packet is shown in Figure 3–34.

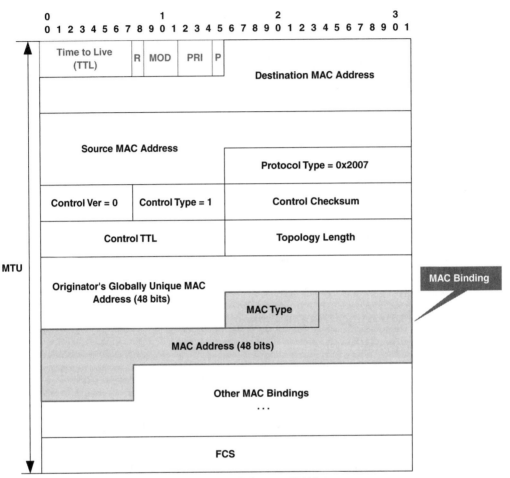

Maximum Transmission Unit (MTU) = max. 9216 bytes

**Figure 3–34**    SRP topology discovery packet format

As can be seen, the topology map starts with the MAC address of the originator, followed by several MAC bindings, each consisting of a MAC Type and a MAC Address field. The MAC Type field is used to denote the ring ID and Wrap status, as shown in Table 3–8.

**Table 3–8**    MAC Type Values Used in Topology Discovery Packets

| BIT | VALUE | USAGE |
|---|---|---|
| 0 | | Reserved |
| 1 | 0<br>1 | Outer Ring<br>Inner Ring |
| 2 | 0<br>1 | Unwrapped<br>Wrapped |
| 3..7 | | Reserved |

### PACKET PROCESSING

Incoming packets are either accepted and passed on to higher layer processes or transited at layer 2 without requiring any layer-3 activity. The MAC logic shown in Figure 3–35 basically consists of a transit buffer, a receive buffer, and a transmit buffer. The MAC logic is responsible for scheduled transmission and implements the SRP fairness algorithm (SRP-fa).

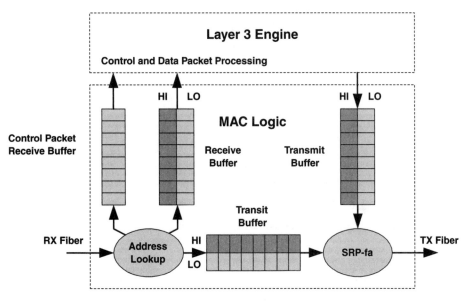

**Figure 3–35**    SRP packet processing

When receiving a packet, the destination address is checked first. If there is a match with the node address, the packet is handed over to the layer-3 engine by placing it into the low- or high-priority receive queue, according to the PRI fields of the packets. Within SRP, unicast traffic is destination stripped; thus, the packet is taken from the ring by the destination and is not put into the transit buffer.

If the destination address of a received packet does not match, the TTL field is decremented. If the TTL field becomes zero, the packet is discarded. Otherwise, the packet is put into the transit buffer for continued circulation. The decision of whether to put the packet into the high- or low-priority queue of the transit buffer is made by examining the PRI field and defined SRP buffer thresholds.

All incoming control packets are processed by the ring node and are stripped from the ring because control packets are always sent p-t-p. Each control packet is analyzed to determine its type by using the Mode field. There are two types of control messages defined. *Locally buffered* control packets are very critical control packets, such as those used for protection switching and are put into a separate receive buffer. *Pass to host* control packets are simply put into the receive buffer used for data packets.

At the transmission side, the packets coming from the layer-3 engine are put into either the low- or high-priority transmit queue, according to the PRI fields of the packets. The scheduler controlled by the SRP-fa then selects a packet out of the low-/high-priority transit buffer or low-/high-priority transmit queue to be sent next.

### Multicasting

SRP combines layer 2 and layer 3, which is the reason for inherent multicasting support. Similar to IP (where class D IP addresses are reserved for multicasting), SRP MAC addresses with the first 3 bytes set to "0x01005E" are reserved for SRP multicast addresses. The least significant bit of the most significant byte, which is called the *multicast bit*, is set to "1" and indicates a multicast packet. SRP supports multicasting for any layer-3 protocol, but the focus is on IP multicast.

Considering IP as the layer-3 protocol, the remaining 3 bytes of the SRP MAC address are used to directly map the 23-bit multicast group ID of the IP multicast address into the SRP multicast address. Figure 3–36 illustrates how the IP multicast address 224.2.175.237 is mapped into an SRP multicast MAC address.

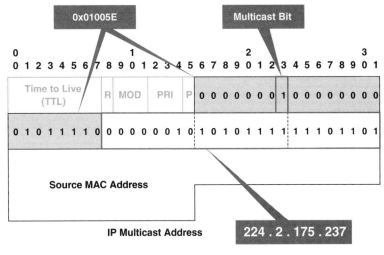

**Figure 3–36**    IP multicast address to SRP multicast address mapping

The ring node originating a multicast packet is responsible for the described mapping procedure. Each node receiving the multicast packet verifies whether it is part of the multicast group. If yes, a copy of the packet is put into the receive buffer. If not, the packet is simply put into the transit buffer for continued circulation around the ring, and the TTL is decremented. The multicast packet is stripped of the ring, either by the source node or by another node, because of TTL expiration.

### Packet Priority

Through packet prioritization, SRP can provide support for real-time applications (video and voice over IP [VoIP]), mission-critical applications, and specific control traffic, which all require stricter delay bounds and jitter constraints. The source node of a packet maps the IP precedence value into the SRP MAC priority. Both prioritization mechanisms use a 3-bit field to denote one of eight priority levels. Thus, the mapping is simply done by copying the three Type of Service bits of the IP header into the 3-bit PRI field of the SRP MAC header.

A configurable priority threshold is used to place the packets either in the low- or the high-priority transit buffer/transmit queue. The scheduler responsible for choosing the next packet to be sent respects packet priority by preferring high-priority packets to low-priority packets and enforces ring packet conservation to avoid discarding of transiting packets.

To facilitate these two rules, the following packet processing hierarchy is used:

1. High-priority transit packets
2. High-priority transmit packets
3. Low-priority transmit packets
4. Low-priority transit packets

In addition, thresholds are applied to the low-priority transit buffer depth to ensure that it does not overflow and that low-priority transit packets do not wait too long behind locally sourced low-priority packets.

When looking at an OC-12c/STM-16c DPT interface with a line rate of 622 Mbps, the low threshold should be set to about 4.4 ms or 320 Kb. The high threshold should be set to about 870 µs higher than the low threshold, thus to about 458 Kb.

### SRP Fairness Algorithm

The SRP-fa is a distributed transmission control algorithm using signaling messages, rate counters, and threshold variables to facilitate the three main goals:

1. By ensuring *global fairness*, each node gets its fair share of ring bandwidth, and other nodes should not be able to create node starvation or excessive delay conditions.
2. *Local optimization* is achieved by using the spatial reuse functionality. Ring nodes can utilize more than their fair share on local ring segments, as long as other nodes are not impeded.
3. *Scalability* makes it possible to build large rings of between 32 and 64 nodes over geographically distributed areas.

Each ring node implements two *rate counters* to assist packet-forwarding decisions on the ring:

- `My_Usage`: Every locally sourced low priority packet is measured with this counter:

```
My_Usage=My_Usage+Packet_Length.
```

High-priority packets are not controlled by the SRP-fa and should be controlled by layer-3 features, such as committed access rate (CAR).

- `Fwd_Rate`: Every packet originated by an upstream node and transiting the local node is measured with this counter:

```
Fwd_Rate=Fwd_Rate+Packet_Length.
```

Both rate counters are also periodically decremented according to a configurable *decay interval,* which allows packet sourcing and packet forwarding credits to accumulate over time and reflect the most recent usage conditions on the ring.

The two implemented threshold variables are:

- `Max_Usage`: A maximum transmission rate for locally sourced traffic, which is static and independent from the SRP-fa and is preconfigured with the `Max_Usage` variable.
- `Allow_Usage`: The node's fair share is represented by the `Allow_Usage` variable. It represents the maximum amount of locally sourced and transmitted low priority traffic. The `Allow_Usage` variable is periodically updated at the above-mentioned decay interval to reflect the current ring traffic conditions.

When a node needs to transmit more traffic than its fair share (`My_Usage > Allow_Usage`), excess packets are buffered in the transmit queue until the `My_Usage` counter decays and is decremented or the `Allow_Usage` variable is updated, as mentioned above.

The example in Figure 3–37 shows how a congested ring node uses fairness signaling to claim its fair share of the ring bandwidth. Node A is transmitting a high amount of traffic to node D. This results in excessive transit traffic for node B, which itself would like to send traffic to node B.

Node B generates fairness signaling messages and sends them upstream toward node A. Node A receives this information, then recalculates its SRP-fa threshold variables and reduces the amount of traffic put on the ring.

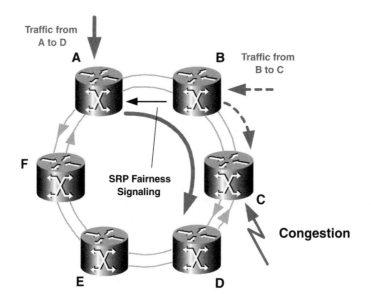

**Figure 3–37**   Node B generates SRP-fa signaling messages because it is congested by node A

In addition, every time the decay interval is over, all ring nodes update their `Allow_Usage` variable and adapt to current ring utilization by using the collected fair share signaling information. Through a comparison with the `Fwd_Rate` counter, the `My_Usage` counter, and the `Allow_Usage` variable, ring nodes calculate their new fair share signaling information. Fair share signaling information is sent out by using the usage packets.

### Intelligent Protection Switching

DPT rings use Intelligent Protection Switching (IPS) to provide powerful self-healing capabilities, which allow the ring to recover automatically from link or node failures by *wrapping* traffic onto the alternate fiber (protection switching). IPS provides functionality analogous to APS for SONET/SDH rings with several important extensions.

IPS is topology knowledge-independent. IPS supports rings with more than 16 nodes on the ring. Furthermore, because DPT statistically multiplexes data packets on the ring, traffic peaks can be handled at up to 100% of the available bandwidth. Within SONET/SDH, protection resources are preallocated in a fixed fashion. Thus, only 50% of the available bandwidth can be used for active traffic.

IPS facilitates fault and performance monitoring. IPS uses ring wrapping to bypass failed nodes or links, which is completely transparent to layer 3. A protection ring event hierarchy is used to prevent partitioning the ring into separate subrings in case of multiple failures.

### IPS MESSAGES

IPS uses IPS control packets for *protection switching signaling*, unlike SONET/ SDH rings, which use the K1 and K2 overhead bytes to implement protection signaling. The IPS control packet format is depicted in Figure 3–38.

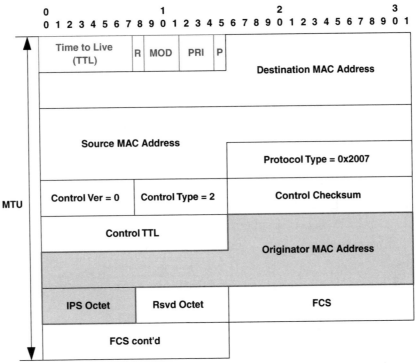

Maximum Transmission Unit (MTU) = max. 9216 bytes

**Figure 3–38**    IPS control packet format

The IPS octet of an IPS message includes information on the IPS Request Type, Wrapping status, and a Path Indicator. Table 3–9 shows details on possible values for the IPS octet. Possible but not listed values are reserved.

**Table 3–9**    IPS Octet Bit Usage

| FIELD | BITS | VALUE | USAGE |
|---|---|---|---|
| IPS Request Type | 0..3 | 1101 | Forced Switch (FS) |
| | | 1011 | Signal Fail (SF) |
| | | 1000 | Signal Degrade (SD) |
| | | 0110 | Manual Switch (MS) |
| | | 0101 | Wait to Restore (WTR) |
| | | 0000 | No Request (I) |
| Path Indicator | 4 | 0 | Short Path (S) |
| | | 1 | Long Path (L) |
| Status Code | 5..7 | 010 | Protection Switch Completed, Traffic Wrapped (W) |
| | | 000 | Idle (I) |

To indicate the contents, IPS messages are quoted in the following format in the upcoming paragraphs:

{Request Type,Source Address,Wrap Status,Path Indicator}.

The possible values for the IPS Request Type, Path Indicator, and Status Code are the abbreviations listed in Table 3–9. The Source Address may be equal to "Src," which is a substitute for the MAC address of the IPS message originating node and "Self," which is a substitute for the node's own MAC address.

SONET/SDH framing is capable of communicating network events, such as fiber cuts or signal degradation, by using its overhead bytes (Table 3–10).

**Table 3–10**   Possible SONET/SDH Network Events

| NAME | DESCRIPTION |
|---|---|
| Loss of Frame (LOF) | detected by monitoring the A1 and A2 byte |
| Loss of Signal (LOS) | in case of a 100-µs-long all-zero pattern |
| Alarm Indication Signal (AIS) | notify a failure to downstream nodes with "1" in bits 6-8 of the K2 byte |
| Bit Error Rate (BER) | counting of parity violations by using the B2 byte |

By monitoring those bytes, **IPS requests** can be initiated. IPS requests are populated with the above-mentioned IPS messages and can be separated into **automatic requests** (initiated by triggering events) and **operator requests** (initiated by CLI commands by the network operator; Tables 3–11 and 3–12).

**Table 3–11**   Automatic IPS Requests

| NAME | DESCRIPTION |
|---|---|
| Signal Fail (SF) | in case of LOS, LOF, BER > SF threshold, … |
| Signal Degrade (SD) | in case of BER > SD threshold |
| Wait to Restore (WTR) | delays unwrapping to prevent IPS oscillations |

**Table 3–12**   Operator IPS Requests

| NAME | DESCRIPTION |
|---|---|
| Lockout of Protection (LOP) | removes active protection switches and prevents subsequent protection switches to control unusual IPS behavior |
| Forced Switch (FS) | forces protecting switching, e.g., while inserting a new node |
| Manual Switch (MS) | like Forced Switch but lower priority |

IPS uses a ***protection request hierarchy*** to handle multiple concurrent requests. Requests with a higher priority can preempt active requests with a lower priority. The request hierarchy, starting with the highest priority, is as follows: Forced Switch (FS), Signal Fail (SF), Signal Degrade (SD), Manual Switch (MS), Wait to Restore (WTR) and Idle (I).

In general, IPS messages can be separated into two groups.

- ***Short path messages*** of the format `{ Req,Src,Stat,S}` are sent over adjacent spans to the node's neighbors and are stripped off by receiving nodes.
- ***Long path messages*** of the format `{ Req,Src,Stat,L}` are sent around the ring and are always forwarded to neighbors until they reach a wrapped node. Long path messages are used to maintain the IPS hierarchy.

### IPS MESSAGE HANDLING STATE MACHINE

IPS messages representing the current ring status are sent p-t-p on both the inner and outer rings. An IPS message-handling state machine is used for processing the messages and controlling the state of the ring node. A node can be in the following three states:

- Idle state: node in normal operation and sending IPS short path messages `{ Idle,Self,I,S}`
- Pass-Through state: node in normal operation, forwarding wrapped traffic and IPS long path messages `{ Req,Src,W,L}`
- Wrapped state: providing protection switching by wrapping traffic to the alternate fiber and sending IPS short path messages `{ Req,Self,W,S}`

Figure 3–39 shows the state transition diagram of the IPS state machine. During normal operation, nodes are in idle state and periodically send out IPS short path messages `{ Idle,Self,I,S}` to indicate their idle state. This can be compared with sending out keepalives.

In case of a local request, the node changes into wrapped state. Through sending out a short path message `{ Req,Self,W,S}`, the node on the other side of the failed span, also called ***mate***, is forced to go into wrapped state, too.

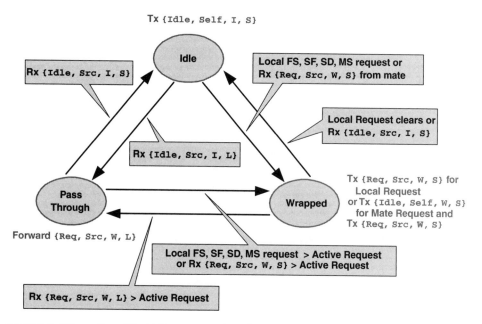

**Figure 3–39**    Simplified IPS message-handling state machine

A node gone into wrapped state because of a local request informs the other nodes on the ring about the ring wrap through sending out the long path message { Req, Self, W, L} . A node gone into wrapped state because of a request received from his mate informs the other nodes on the ring through sending out the long path message { Idle, Self, W, L} .

The nodes on the ring receiving these long path messages { Req, Src, W, L} automatically go into pass-through state and forward the received long path messages.

In case of second concurrent failure event with a higher priority, the affected nodes in pass-through state go into wrapped state and force nodes in wrapped state to change to pass-through state by sending short and long path signaling messages.

The IPS message flow for both the normal operation and a failure of the inner ring fiber connection between routers A and B is shown in Figure 3–40.

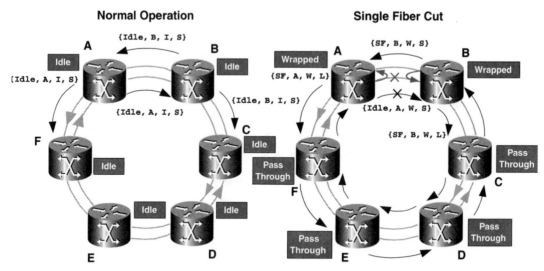

**Figure 3–40**    IPS message flow during normal operation and in case of a single fiber cut

## Node Pass-Through Mode

A ring node may have limited layer-3 functionality because of manual intervention by the operator or because of a software crash. For such situations, the node may go into node pass-through mode. The node pass-through mode is different than the pass-through status described in the IPS section. In the node pass-through mode, the node does not put any packets onto the ring. It simply forwards packets around the ring without modifying them. The node also does not participate in any control packet exchange. Instead, it just behaves like a signal regenerator.

### Network Management

DPT products use the SONET Management Information Base (MIB) as defined in RFC 2558 [IETF-9] and the SRP MIB [IETF-10] for current alarm status and performance monitoring information, as well as historical performance monitoring information.

### Clocking and Synchronization

SRP ring nodes operate in free running mode. This means that the receive clock is derived from the received data stream. The transmit clock is derived from a local oscillator with an accuracy within 20 ppm. This eliminates the need for expensive clock synchronization, as is required in SONET/SDH networks.

The small differences in clock frequencies between nodes are handled by inserting small amounts of idle bandwidth at each node's output.

## Data Transmission Technology Comparison

The following section highlights the differences between the IP transport alternatives: IP over ATM, POS, or DPT. The following comparison analyzes the transmission efficiency for an OC-3/STM-1 interface with its transmission rate of 155 Mbps for the different transport alternatives listed above.

### Encapsulation Overhead

When carrying IP traffic, which is not delay- and jitter-sensitive, ATM uses the AAL5 for encapsulating IP packets. The encapsulated packets are split into ATM cells with 48-bit payload by the segmentation and reassembly (SAR) layer.

$$AAL5\_PDU = IPSIZE + SNAP\_HD + AAL5\_OH$$
$$= IPSIZE + 16 \tag{3-3}$$

$$\frac{Cells}{Packet} = \frac{AAL5\_PDU}{48} \tag{3-4}$$

$$ATM\_PSIZE = \frac{Cells}{Packet} * 53 = roundup(\frac{IPSIZE + 16}{48}) * 53 \tag{3-5}$$

where AAL5_PDU = ATM AAL5 protocol data unit; AAL5_OH = ATM AAL5 overhead; ATM_PSIZE = ATM packet size; IPSIZE = IP packet size; and SNAP_HD = Subnetwork Access Protocol header.

POS uses PPP as the IP packet encapsulation mechanism with HDLC-like framing.

$$POS\_PISZE = IPSIZE + POS\_OH = IPSIZE + 7 \text{, for CRC16} \quad (3\text{-}6)$$

$$POS\_PISZE = IPSIZE + POS\_OH = IPSIZE + 9 \text{, for CRC32} \quad (3\text{-}7)$$

where POS_PSIZE = POS packet size; POS_OH = POS overhead; and CRC = cyclic redundancy check.

DPT uses the SRP MAC layer protocol with its SRP MAC header similar to the Ethernet MAC header for carrying IP traffic.

$$DPT\_PSIZE = IPSIZE + DPT\_OH = IPSIZE + 23 \tag{3-8}$$

where `DPT_PSIZE` = DPT packet size and `DPT_OH` = DPT overhead.

### Framing Overhead

These packets of the size derived in Equations (3–5), (3–6)/(3–7), and (3–8) are then mapped into SONET/SDH frames.

$$FSIZE = rows * columns = 9 * 270 = 2430\, bytes \tag{3-9}$$

$$PLD = FSIZE - SONET/SDH\_OH$$
$$= 2430 - 9 * (9 + 1) = 2340\, bytes \tag{3-10}$$

where `FSIZE` = frame size; `PLD` = payload; and `SONET/SDH_OH` = SONET/SDH overhead.

A SONET/SDH 155-Mbps interface delivers 8,000 frames per second and is capable of carrying ATM, POS, or DPT data stream packets. The maximum packet rate, throughput, and maximum transmission bit rate are derived through Equations (3–11), (3–12) and (3–13).

$$PacketRate = \frac{PLD/sec}{PSIZE} = \frac{2340 * 8000}{PSIZE} \tag{3-11}$$

$$ThroughPut = PacketRate * IPSIZE \tag{3-12}$$

$$TransmissionRate = ThroughPut * 8 \tag{3-13}$$

where `PSIZE` = packet size (ATM, DPT, or POS)

The encapsulation and framing efficiency of each technology results in

$$Efficiency = \frac{ThroughPut}{PLD/sec} = \frac{PacketRate * IPSIZE}{2340 * 8000} \tag{3-14}$$

Considering the SONET/SDH overhead, the total efficiency can be derived through

$$TotalEfficiency = Efficiency * \frac{2340}{2430} = \frac{PacketRate * IPSIZE}{2340 * 8000} * \frac{2340}{2430}$$
$$= \frac{PacketRate * IPSIZE}{8000 * 2430} \tag{3-15}$$

Tables 3–13, 3–14, 3–15, and 3–16 show the results of the equations discussed above when varying the IP packet size in the range of 46 bytes up to 4470 bytes.

**Table 3–13**    ATM Encapsulation and Framing Characteristics

| IP SIZE [BYTES] | CELLS/ PACKET | ATM_P SIZE [BYTES] | PACKET RATE [PPS] | THROUGHPUT [MB/SEC] | TRANSMISSION RATE [MBPS] | EFFICIENCY [%] | TOTAL EFFICIENCY [%] |
|---|---|---|---|---|---|---|---|
| 46 | 2 | 106 | 176603 | 8.124 | 64.990 | 43.40% | 41.79% |
| 110 | 3 | 159 | 117735 | 12.951 | 103.607 | 69.18% | 66.62% |
| 238 | 6 | 318 | 58867 | 14.010 | 112.083 | 74.84% | 72.07% |
| 494 | 11 | 583 | 32109 | 15.862 | 126.895 | 84.73% | 81.59% |
| 1006 | 22 | 1166 | 16054 | 16.150 | 129.203 | 86.27% | 83.08% |
| 1500 | 32 | 1696 | 11037 | 16.556 | 132.444 | 88.44% | 85.16% |
| 2030 | 43 | 2279 | 8214 | 16.674 | 133.395 | 89.07% | 85.77% |
| 4334 | 91 | 4823 | 3881 | 16.820 | 134.562 | 89.85% | 86.52% |
| 4470 | 94 | 4982 | 3757 | 16.794 | 134.350 | 89.71% | 86.39% |

**Table 3–14**    POS Encapsulation and Framing Characteristics (CRC16)

| IP SIZE [BYTES] | POS_P SIZE [BYTES] | PACKET RATE [PPS] | THROUGHPUT [MB/SEC] | TRANSMISSION RATE [MBPS] | EFFICIENCY [%] | TOTAL EFFICIENCY [%] |
|---|---|---|---|---|---|---|
| 46 | 53 | 353207 | 16.248 | 129.980 | 86.79% | 83.58% |
| 110 | 117 | 160000 | 17.600 | 140.800 | 94.02% | 90.53% |
| 238 | 245 | 76408 | 18.185 | 145.481 | 97.14% | 93.54% |
| 494 | 501 | 37365 | 18.458 | 147.666 | 98.60% | 94.95% |
| 1006 | 1013 | 18479 | 18.590 | 148.719 | 99.30% | 95.63% |
| 1500 | 1507 | 12422 | 18.633 | 149.064 | 99.54% | 95.85% |
| 2030 | 2037 | 9189 | 18.654 | 149.229 | 99.65% | 95.96% |
| 4334 | 4341 | 4312 | 18.688 | 149.506 | 99.83% | 96.13% |
| 4470 | 4477 | 4181 | 18.689 | 149.513 | 99.83% | 96.14% |

**Table 3–15**    POS Encapsulation and Framing Characteristics (CRC32)

| IP SIZE [BYTES] | POS_P SIZE [BYTES] | PACKET RATE [PPS] | THROUGHPUT [MB/SEC] | TRANSMISSION RATE [MBPS] | EFFICIENCY [%] | TOTAL EFFICIENCY [%] |
|---|---|---|---|---|---|---|
| 46 | 55 | 340363 | 15.657 | 125.254 | 83.64% | 80.54% |
| 110 | 119 | 157310 | 17.304 | 138.433 | 92.44% | 89.01% |
| 238 | 247 | 75789 | 18.038 | 144.302 | 96.36% | 92.79% |
| 494 | 503 | 37216 | 18.385 | 147.078 | 98.21% | 94.57% |
| 1006 | 1015 | 18443 | 18.554 | 148.429 | 99.11% | 95.44% |
| 1500 | 1509 | 12405 | 18.608 | 148.860 | 99.40% | 95.72% |
| 2030 | 2039 | 9180 | 18.635 | 149.083 | 99.55% | 95.86% |
| 4334 | 4343 | 4310 | 18.680 | 149.436 | 99.78% | 96.09% |
| 4470 | 4479 | 4179 | 18.680 | 149.441 | 99.79% | 96.09% |

**Table 3–16**    DPT Encapsulation and Framing Characteristics

| IP SIZE [BYTES] | DPT_P SIZE [BYTES] | PACKET RATE [PPS] | THROUGHPUT [MB/SEC] | TRANSMISSION RATE [MBPS] | EFFICIENCY [%] | TOTAL EFFICIENCY [%] |
|---|---|---|---|---|---|---|
| 46 | 69 | 271304 | 12.480 | 99.840 | 66.67% | 64.20% |
| 110 | 133 | 140751 | 15.483 | 123.861 | 82.71% | 79.64% |
| 238 | 261 | 71724 | 17.070 | 136.562 | 91.19% | 87.81% |
| 494 | 517 | 36208 | 17.887 | 143.094 | 95.55% | 92.01% |
| 1006 | 1029 | 18192 | 18.301 | 146.409 | 97.76% | 94.14% |
| 1500 | 1523 | 12291 | 18.437 | 147.492 | 98.49% | 94.84% |
| 2030 | 2053 | 9118 | 18.510 | 148.076 | 98.88% | 95.21% |
| 4334 | 4357 | 4296 | 18.619 | 148.951 | 99.46% | 95.78% |
| 4470 | 4493 | 4166 | 18.622 | 148.976 | 99.48% | 95.79% |

Figure 3–41 and Figure 3–42 illustrate the two most important characteristics, the maximum possible transmission rate, and the total efficiency as a function of the IP packet size.

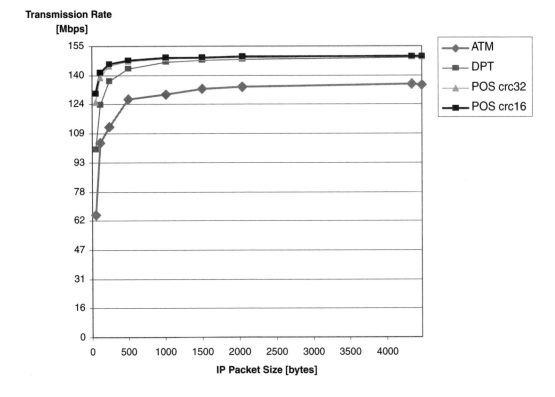

**Figure 3–41**    IP transmission rate for an OC-3/STM-1 interface as a function of the IP packet size

**Total Efficiency [%]**

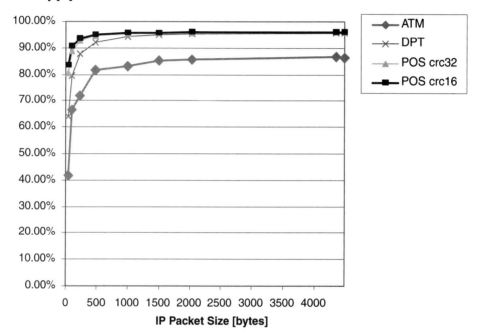

**Figure 3–42**   Total transmission efficiency as a function of the IP packet size

These diagrams show a broad IP packet size range. Every application uses a variety of different packet sizes. Typical applications use packets with sizes between 50 and 1,000 bytes. The minority of applications (such as file transfers) use big packets of about 3,000–4,000 bytes.

Assuming applications using IP packets of about 50–1,000 bytes, the OC-3/ STM-1 line rate of 155 Mbps cannot be reached by any of these three technologies. But, when comparing ATM to the "optical networking technologies" POS and DPT, a big difference can be seen. The efficiency when using ATM for transporting packets up to 1,000 bytes ranges at about 40–80%. POS and DPT achieve efficiency at about 60–95% in that region. This big difference, because of AAL5 and SAR layer overhead functions, is a big disadvantage of ATM when implementing efficient high-speed backbones.

When comparing POS and DPT, both achieve efficiency at about 95% when transmitting big IP packets. However, DPT is about 20% less efficient than

POS when transmitting 46-byte IP packets. The reason is the Ethernet-like framing of DPT. Addressing, priority, and control functions for the MAC layer require a MAC header bigger than the PPP header used by POS.

## MPLS Traffic Engineering (MPLS-TE)

The concept of MPLS has already been described in Chapter 1, "Introduction to Carrier Network Architectures." MPLS delivers the ability to perform traffic engineering at the IP layer. In the following section, the concept and mechanisms behind the MPLS-TE application are described in order to provide the necessary background to understand the concept of the Multiprotocol Lambda Switching (MPLmS) architecture, which is developed to provide dynamic wavelength provisioning in the OTN.

### MPLS-TE Architecture

#### WHY MPLS-TE
Standard routing protocols compute the optimum path from a source to a certain destination, considering a routing metric such as hop-count, cost, or link bandwidth. As a result, a single least-cost path is chosen. Although there might be an alternate path, only the one determined by the routing protocol is used to carry traffic. This leads to an inefficient utilization of network resources.

The example shown in Figure 3–43 illustrates this situation. A standard routing protocol would route traffic from router A to the routers B and C across the path A-W-Z-B and A-W-Z-C. This may lead to congestion at the link between the routers W and Z, although the links between W, X, and Z may be underutilized.

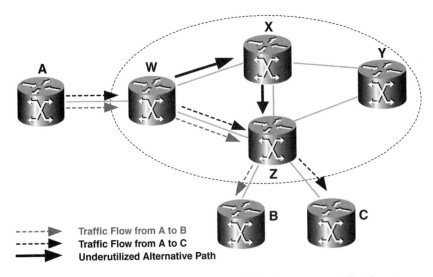

**Figure 3–43**   Standard routing leading to inefficient resource utilization

An approach for resolving this problem is to manipulate the Interior Gateway Protocol (IGP) metrics. In the example shown in Figure 3–43, increasing the metric for the link W-Z would switch both traffic flows from router A to B and from router A to C onto the paths A-W-X-Z-B and A-W-X-Z-C. Now the link W-Z might be underutilized.

Policy-based routing has been the next step for optimizing network resource utilization. Through defining complex access lists, certain traffic flows could be characterized. By configuring these access lists on all the routers throughout the whole network, traffic could be routed on a per-flow basis.

Both IGP metric manipulation and policy-based routing are static solutions and do not scale. Traffic engineering provides a dynamic and efficient solution for this problem.

MPLS-TE provides traffic-driven IGP route calculation functionality on a per-traffic trunk basis. Other than with standard routing protocols, which are only topology driven, utilization of network resources and resilience attributes are analyzed and taken into account during path computation. Multiple paths are possible from a source to a destination, and the best path, according to the current network situation, is taken to ensure optimum network utilization. Network optimization or reoptimization may occur automatically or off-line, using a centralized traffic engineering server.

With MPLS-TE, the situation discussed before can be easily solved, as shown in Figure 3–44. Two traffic trunks are established—one from label switch router (LSR) A to LSR B and one from LSR A to LSR C. Suppose that the path for the traffic trunk from A to B is calculated first and the path A-W-Z-B is selected. During the next step, the path for the other traffic trunk is calculated. It might be detected that the link W-Z has, for example, only half of the required bandwidth left available, but the links W-X and X-Z have more than twice of the bandwidth requested by the traffic trunk available. As a consequence, the path A-W-X-Z-C is chosen, and optimized network utilization is achieved.

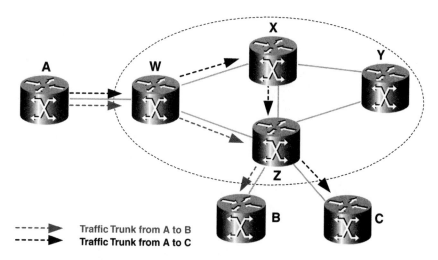

**Figure 3–44**    Route traffic trunks with traffic engineering according to the accurate network utilization

### MPLS-TE COMPONENTS
MPLS-TE incorporates a couple of functions. The MPLS-TE functional components and their relationships are shown in Figure 3–45.

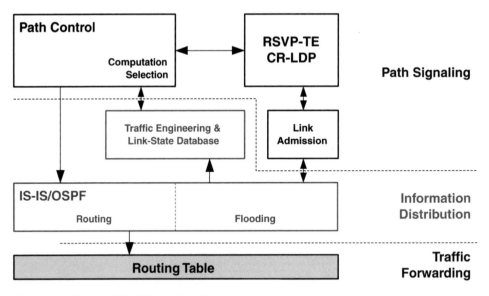

**Figure 3–45**    MPLS-TE functional components

The routing protocol is a central part in the traffic engineering architecture. Because all nodes in the network must know the whole topology of the network to be able to compute a path, only a link-state protocol can be used. This standard IGP, such as OSPF or IS-IS, must be extended with some traffic engineering extensions to enable the routing protocol to carry resource and policy information of the network. The two IETF drafts "Traffic Engineering Extensions to OSPF" [IETF-17] and "IS-IS Extensions for Traffic Engineering," [IETF-23] propose those extensions.

The collected information is used to maintain, on the one hand, a link-state database providing a topological view of the whole network and, on the other hand, a traffic engineering database (TE database) storing resource and link utilization information. These databases are consolidated by the path control component. A constrained-based routing algorithm, also called *Constrained Shortest Path First* (CSPF), is used to compute the best path according to the information collected in the two databases.

A signaling protocol such as Resource Reservation Protocol (RSVP) or LDP is used to set up a label-switched path (LSP) along the selected path through the network. Also, these protocols have to be extended to enable LSP setup and maintenance functionality. These changes are specified in the two IETF drafts

"RSVP-TE: Extensions to RSVP for LSP Tunnels" [IETF-24] and "Constraint-Based LSP Setup Using LDP" [IETF-25]. During the LSP setup, each node has to check whether the requested bandwidth is available. The link admission control component acts as an interface between the IGP collecting the resource information and the LSP signaling component requesting the bandwidth. If bandwidth is available, it is allocated. If not, an active LSP might be preempted or the LSP setup fails.

The established LSPs are acting as links, are fed back into the routing protocol, and are used as next hop for traffic going to the tail end of the LSP or to destinations behind the tail end. Standard MPLS is used as the forwarding mechanism.

### DEFINING THE TERM *TRAFFIC TRUNK*

Traffic engineering introduces the term ***traffic trunk***. A traffic trunk is considered to be an aggregation of data flows following the same path through the network and belonging to the same Class of Service (CoS).

Thus, at an ingress node of the network, for each CoS, all the traffic to the same egress node is combined and transported over one LSP. By doing this, traffic flows to be handled are reduced to a number of traffic trunks traversed through LSPs.

The terms ***traffic trunk*** and ***LSP*** are often used synonymously but there must be differentiation. The traffic trunk is an abstract representation of traffic with specific characteristics assigned, and the LSP is a specification of a label switched path that the traffic traverses through. Thus, a traffic trunk is completely separated from the LSP and might also use another LSP fulfilling the trunk requirements.

How these LSPs are to be routed regarding their QoS and resilience requirements is controlled by assigning trunk attributes, which are then considered during CSPF computation and link admission.

It is important to note that traffic trunks are unidirectional. Thus, bidirectional inter-POP traffic flow is handled by defining two traffic trunks and setting up two LSPs in opposite directions.

## MPLS-TE Attributes

### TRUNK ATTRIBUTES

At the head-end LSR, several trunk attributes are assigned to each traffic trunk to specify the characteristics of the traffic trunk.

- Bandwidth
- Resource class affinity
- Path selection policy
- Priority/preemption
- Adaptability
- Resilience

### *BANDWIDTH*

The bandwidth attribute specifies the amount of bandwidth used by the traffic trunk.

### *RESOURCE CLASS AFFINITY*

The resource class affinity attribute provides the ability explicitly to include or exclude some links/resources of the network during path CSPF computation.

The resource class affinity attribute consists of a 32-bit resource class affinity string and a 32-bit resource class mask. During the resource class policy check, the trunk's resource class affinity string is AND, combined with the resource class mask. The result is XOR, combined with the resource class string of the link. If the final result is "all zero," the link has passed the policy check and can be used by the trunk.

The default for the resource class string might be defined as "0x0000." A default mask of "0x00FF" means that, without any explicit configuration, the policy check does not care about the upper half of the resource class string.

### *PATH SELECTION POLICY*

The path selection policy defines in which order the head-end LSR should select explicit paths. The explicit paths can be manually configured either by defining the hops following from the ingress to the egress LSR or by dynamically computing them, using the CSPF algorithm.

### PRIORITY/PREEMPTION

The priority or preemption attribute characterizes the importance of a traffic trunk. Traffic trunks carrying critical traffic, such as voice or video, must be prioritized over traffic trunks carrying best-effort Internet traffic. Also, in situations of network contention or network failures, the order in which traffic trunks are to be routed is very important.

The attribute consists of two parameters. The reservation priority (also called *hold priority*) defines whether a traffic trunk is allowed to keep its existing reservation when another traffic trunk is attempting to take away the reserved resources. The preemption priority or setup priority is used to give a traffic trunk the ability to claim already reserved resources from an existing traffic trunk.

According to the priority mechanism defined by IP Type of Service (TOS), eight setup and hold priorities are available, with 0 used as the highest and 7 as the lowest priority.

### ADAPTABILITY

The adaptability attribute is used to specify whether the traffic trunk should be reoptimized to adapt to changing network characteristics. It is a binary parameter with the two possible values "enable reoptimization" or "disable reoptimization." Reoptimization can be done administratively or dynamically via periodic recalculation at the head-end LSR of the LSPs.

The rerouting procedure must not affect the active LSP for the traffic trunk. First, a new LSP is established, then the traffic trunk is switched onto this new trunk. To avoid that during this procedure, twice the bandwidth is reserved or, even worse, if the second LSP cannot be established, the "shared explicit" principle is applied. That means that no additional bandwidth is reserved but taken from the LSP to be reoptimized when the switchover is done.

### RESILIENCE

The resilience attribute is used for restoration purposes. It specifies how to react to situations where the LSP of the traffic trunk does not exist anymore, due to network failures or preemption.

Several restoration policies can be considered. Restoration might be disabled in the case where additional restoration mechanisms are available. A traffic trunk might be rerouted through an alternate LSP, whereas the alternate path can be preestablished (hot-standby) or can also be established on demand when the failure occurs.

### RESOURCE ATTRIBUTES

The resource attributes are flooded by the IGP, collected in the TE database, and used to constrain the placement of traffic trunks.

- Available bandwidth
- Resource class
- Administrative weight

#### *AVAILABLE BANDWIDTH*

The available bandwidth attribute specifies the bandwidth provided on a link per setup priority. Because it might be desired to oversubscribe or underutilize a certain link, the maximum allocation multiplier (MAM) is used as a scaling factor to set the value for the available bandwidth different from the actual link bandwidth.

#### *RESOURCE CLASS*

The resource class attribute is used to assign a link to a certain class of resources. This 32-bit string is used during the path setup procedure to determine whether to route or not to route the LSP over the link. The exact description of the policy check has already been discussed at the trunk attribute description.

#### *ADMINISTRATIVE WEIGHT*

This attribute is used during traffic engineering calculations. The default value equals the administrative weight of the IGP, but there could possibly be two different values used. One value is used for IGP SPF calculation and one for routing LSPs across the IGP domain.

## Information Distribution

Trunk attributes are configured locally at the head-end LSRs, and resource attributes are configured locally at the LSR for each link. This information must be spread across the network so that each head-end LSR has all the necessary information to be able to compute paths for traffic trunks. Resource information is distributed using extended IGP flooding mechanisms. Because flooding of updates carrying the defined resource attributes generates a significant amount of traffic sent throughout the network, it must be controlled and optimized.

First, each node in the network checks its resource state on a specified timer interval and initiates a flooding process if it is different from the last distributed state. Second, flooding is done on a significant change, where a "significant change" means that a threshold has been crossed. There are multiple up- and down-thresholds defined, and each time one of them is crossed, IGP updates are sent out. Due to the threshold scheme, the resource information each node has is not 100% exact throughout the network. For example, node A might want to be able to set up an LSP over node B, but node B has already reserved some additional bandwidth without crossing the defined threshold. As a result, the LSP setup fails. This event acts as a trigger for node B to flood its accurate resource state.

### BANDWIDTH ACCOUNTING

The IGP does not just distribute the maximum available link bandwidth. The more interesting thing for traffic engineering is the amount of bandwidth left available for each of the eight priorities. Figure 3–46 shows an example how the IGP keeps track of the ongoing bandwidth reservations.

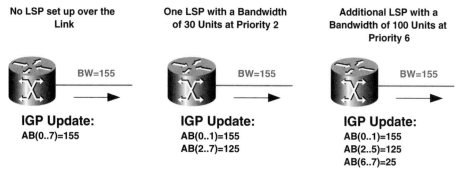

**Figure 3–46**    IGP flooding of the available bandwidth of a link

Each link has a maximum link bandwidth specified. In the example shown, there are 155 units available on the link. After the first LSP is set up at setup priority 2 with an amount of 30 bandwidth units, the IGP distributes 155 units available for important LSPs having a priority of 0 or 1 and only 125 units for all other LSPs.

Setting up another LSP with 100 bandwidth units at setup priority 6 results in IGP updates advertising 155 units available for priorities 0 and 1, 125 units available for priorities 2–5, and only 25 units available for priorities 6 and 7.

In summary, there are now 125 units reserved, thus, only 25 units left. However, considering another setup request, e.g., for an LSP with 60 units at setup priority 1, the 100-unit LSP at priority 6 would be torn down, freeing up enough bandwidth for the requested high-priority LSP.

**OSPF EXTENSIONS**
The extensions to OSPF necessary to support traffic engineering have been proposed in the IETF draft "Traffic Engineering Extensions to OSP" [IETF-17]. Prior to this IETF draft, opaque link state advertisements (LSAs) were defined in RFC 2370 "The OSPF Opaque LSA Option" [IETF-26] to be able to adapt OSPF to future requirements.

The defined TE extensions make use of these opaque LSA definitions. For describing LSRs, p-t-p connections, and connections to multiaccess networks, a new LSA called the *Traffic Engineering LSA* is defined. This LSA is an opaque LSA of type 10, which has an area-wide flooding scope. The Link State ID consists of an 8-bit-long Type value, which is set to 1 and 24 bits of type-specific data. These 24 bits are split into 8 reserved bits and a 16-bit-long instance field. The instance value is used to handle multiple LSAs per LSR. The LSA format is shown in Figure 3–47.

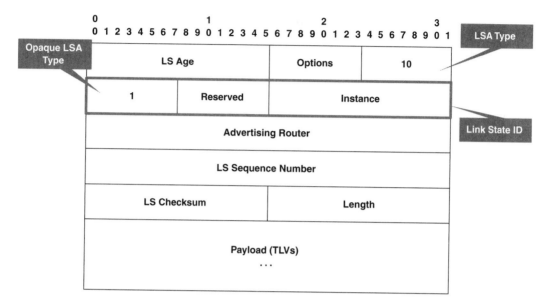

**Figure 3–47**    Traffic Engineering LSA format used by OSPF Traffic Engineering

Multiaccess links are described using the already existing network LSA. A network LSA is generated for every transit broadcast or nonbroadcast multi-access (NBMA) network and contains a list of all LSRs attached to the network. The Link State ID field contains the interface IP address of the designated LSR of the NBMA network. The resulting LSA format is shown in Figure 3–48.

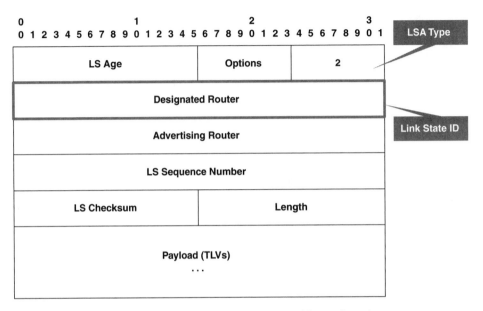

**Figure 3–48**    Network LSA format used by OSPF Traffic Engineering

Both LSAs have some fields in common. The ***LS Age*** field denotes the age of an LSA as an unsigned 16-bit integer value. The LS Age measured in seconds is incremented at each LSR as it is traveling across the network and also when it is held in the LSR's database. LSRs receiving two identical LSAs with both equal LS Sequence Numbers and LS Checksums examine the LS Age field, and the LSA with the smaller LS Age value is chosen as the more recent and valid one.

The ***Options*** field is used to define certain capabilities of the LSR and to communicate them by exchanging LSAs. By using the Option field, it is possible to exclude LSRs with reduced capabilities during routing table calculation and to forward traffic around them. The optional OSPF capability function is implemented within not only LSAs, but also within Hello packets and database description packets.

The ***Advertising Router*** field contains the LSR ID of the LSA-originating LSR.

The ***Sequence Number*** is used to differentiate duplicate LSAs, to detect the old one, and to discard it. The Sequence Number is a signed 32-bit integer value, whereas a higher sequence number value denotes a more recent LSA.

The **LS Checksum** field contains a checksum applied over the whole LSA (except the LS Age field) to detect data corruption of an LSA. The LS Age field is not included in the checksum calculation to enable LSA aging without checksum recalculation.

The **Length** field denotes the length of the LSA in bytes.

The payload of both LSAs contains one or more Type/Length/Value (TLV) objects. There are two top-level TLVs defined: the Router Address and Link TLV. Within these two TLVs, several sub-TLVs are nested, specifying parameters such as link type, bandwidth, or IP address.

Both the top-level TLVs and sub-TLVs are 32-bit aligned and have the format shown in Figure 3–49. A 16-bit field specifies the TLV type. The 16-bit length field specifies the number of bytes contained in the value portion.

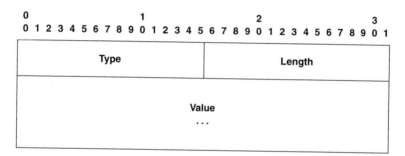

**Figure 3–49**    TLV and sub-TLV format used by OSPF Traffic Engineering

### ROUTER ADDRESS TLV

The Router Address TLV specifies an IP address of the advertising LSR. Normally, the IP address of a logical loopback interface is used, because it is always active, as long as the LSR is alive. The Router Address TLV has the TLV type 1 and a fixed length of 4 bytes.

### LINK TLV

The Link TLV describes the characteristics of a single link. It contains a number of sub-TLVs, each specifying a certain parameter required for traffic engineering. The Link TLV has the TLV type 2 and a variable length, depending on the number of nested sub-TLVs.

The **Link Type sub-TLV** defines whether the link is a p-t-p or multiaccess link.

The *Link ID sub-TLV* specifies the remote side of the link. For p-t-p links, this is the LSR ID of the neighbor. For multiaccess links, this is the LSR ID of the designated router of the broadcast or nonbroadcast network.

The *Local Interface IP Address sub-TLV* and *Remote Interface IP Address sub-TLV* specify the interface IP addresses corresponding to the link at both sides.

The *Traffic Engineering Metric sub-TLV* defines the link metric used for traffic engineering. This metric can be different to the standard OSPF metric.

The *Maximum Bandwidth sub-TLV* specifies the true link bandwidth available in the direction from the LSA-originating LSR to its neighbor. The value is stated according to the Institute of Electrical and Electronics Engineering (IEEE) floating-point format in bytes per second.

The *Maximum Reservable Bandwidth sub-TLV* specifies the maximum bandwidth that might be reserved on the link in the direction from the LSA-originating LSR to its neighbor. Link oversubscription is possible because the value can be set greater than the value specified by the Maximum Bandwidth sub-TLV. Also, this value is stated according to the IEEE floating-point format in bps.

The amount of bandwidth not yet reserved at each of the priority levels is denoted by the *Unreserved Bandwidth sub-TLV.* The value is stated according to the IEEE floating-point format in bps and is equal or less than the value specified by the Maximum Reservable Bandwidth sub-TLV.

The *Resource Class sub-TLV* specifies the Resource Class membership of the link, which is used during path computation for explicit including or excluding of certain links in the network.

### IS-IS EXTENSIONS

The extensions to IS-IS have been proposed in the IETF draft "IS-IS Extensions for Traffic Engineering" [IETF-23] and specify two new TLVs replacing the existing Neighbor TLV and IP Reachability TLV. With these new TLVs, information about the link characteristics can be carried within the LSAs. In addition, the metric space for IP prefixes has been enhanced to provide more flexibility.

As opposed to OSPF, the standard format of the IS-IS LSAs can be used. The new TLVs are carried in the standard Link State Protocol data units as before. The format of the TLVs is the same as with OSPF shown in Figure 3–49.

### ROUTER ID TLV

The already existing Router ID TLV is used to specify the IP address of the advertising LSR. The Router ID TLV has TLV type 134 and can be compared to the Router Address TLV of OSPF.

### EXTENDED IP REACHABILITY TLV

This TLV does not provide information and functionality directly required for traffic engineering. It defines a 32-bit metric opposed to the standard 6-bit metric and introduces a control bit to avoid routing loops that emerge through prefix redistribution between level 1 and level 2 areas. The Extended IP Reachability TLV has TLV type 135 assigned.

### EXTENDED IS REACHABILITY TLV

The Extended IS Reachability TLV can be compared with the Link TLV of OSPF. It contains information about IS neighbors and specifies the appropriate link characteristics with its nested sub-TLVs. The Extended IS Reachability TLV has TLV type 22 assigned.

When comparing the list of proposed sub-TLVs with the one of the OSPF extensions, it can be seen that both the OSPF and IS-IS extensions reflect similar information elements and functions.

The *IPv4 Interface Address sub-TLV* specifies the local link IP address of the LSR, and the *IPv4 Neighbor Address sub-TLV* specifies the link IP address of the remote side.

The *Traffic Engineering Default Metric sub-TLV* is used to define a metric for traffic engineering, which is different than the metric used by the standard IS-IS routing mechanism.

The *Maximum Link Bandwidth sub-TLV* specifies the true link bandwidth using a 32-bit value in the IEEE floating-point format.

To allow link oversubscription, the *Maximum Reservable Bandwidth sub-TLV* defines how much bandwidth can be allocated in the direction from the LSR to the remote LSR, using a 32-bit value in the IEEE floating-point format.

The unallocated bandwidth at each of the eight priority levels is carried in the *Unreserved Bandwidth sub-TLV,* using a 32-bit value in IEEE floating-point format.

The *Administrative Group sub-TLV* specifies the Resource Class membership as is done by the Resource Class sub-TLV of OSPF.

## Path Computation and Selection

Path computation and selection for the LSP of a traffic trunk is done at the head-end LSR. It uses the topology and resource information collected through the extended IGP and computes a suitable path through the network to the tail end of the traffic trunk.

The computed path satisfies a set of requirements assigned to the traffic trunk. These constraints are imposed by administrative policies such as resource class affinity or priority and the actual state of the network, typically depending on the topology and resource availability. Computation of such a path is referred to as *constrained-based routing*. An extended standard Shortest Path First (SPF) algorithm supporting this functionality is known as the *Constrained-Based SPF* (CSPF) algorithm.

Path computation is performed if a new trunk has to be set up, if an existing trunk goes down because of a network failure, or if an existing trunk should be rerouted to reoptimize the network.

There are several input variables used by the CSPF algorithm: First are the configured traffic trunk attributes; second, the attributes associated with the resources; third, the topology information. Both the resource attributes and the topology information are available through the IGP.

In the first step, the CSPF algorithm deletes all links providing insufficient link resources. In the second step, the resource class policy check is performed, and all links with an invalid resource class attribute are also pruned.

In the third step, the shortest distance path is computed. First, the path with the biggest leftover bandwidth is selected, then the path with the smallest hop-count is selected.

The output of the CSPF algorithm is an explicit route specifying the sequence of all hops from the head end to the tail end of the LSP. Each hop is denoted by a LSR IP address, which can be either the interface IP address of the link or a loopback IP address of the LSR, if unnumbered links are used.

Figure 3–50 shows a path computation example for a traffic trunk to be set up from LSR A to LSR B. The trunk's requested bandwidth is 40 units at a setup priority of 5. The resource class of the trunk is specified with a resource class affinity string set to "0x0000" and a resource class mask set to "0x00FF."

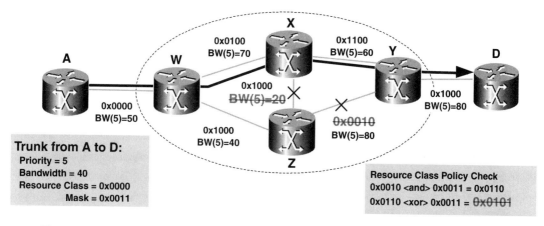

**Figure 3–50**   Constrained-based path computation example

As denoted in the figure, the link between LSRs X and Z has less bandwidth available, and the link between LSRs Z and Y fails to pass the resource class policy check.

Path computation is very CPU-intensive and, therefore, might also be considered to be performed centrally on a traffic engineering server. Especially for network reoptimization purposes, a centralized solution for off-line computation is very useful.

### LSP TUNNEL SETUP

Path setup is done through a signaling protocol that takes the explicit path as input and requests all the LSRs along that path to assign a label for an LSP to carry the requested traffic trunk. Signaling protocols LDP or RSVP might be used; however, RSVP seems to have become the dominating solution. Therefore, the main focus in this book is placed on RSVP.

To support the required functions for traffic engineering, RSVP has been extended. These extensions have been proposed in the IETF draft "RSVP-TE: Extensions to RSVP for LSP Tunnels" [IETF-24], which provides the background for this section.

#### LSP SETUP PROCEDURE

Both the LSP setup and the path along which the LSP has to go through the network are controlled at the head-end LSR of the traffic trunk. As a result, the LSP can be handled as a tunnel; thus, the LSP is commonly referred to as the

*LSP tunnel.* Figure 3–51 illustrates the setup procedure using RSVP for an LSP from LSR A to D.

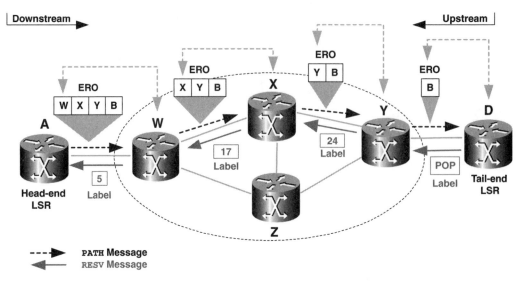

**Figure 3–51**    LSP setup example using RSVP

*LSP SETUP REQUEST*

RSVP uses downstream-on-demand label distribution. The head-end LSR is initiating the LSP setup by sending a PATH message to the next hop LSR of the explicit path. The PATH message includes a couple of objects carrying the necessary information for LSP setup.

The LABEL_REQUEST object denotes that label binding is desired and also specifies the network layer protocol, which might not necessarily be IP.

The EXPLICIT_ROUTE object (ERO) contains the computed path for the LSP, represented by a list of all hops along the path. A hop can be denoted through an IP prefix or autonomous system (AS) number.

The RECORD_ROUTE object is used to keep track of the actual path the LSP traverses.

The SESSION_ATTRIBUTE object is used for session identification and for carrying control information such as preemption/priority, local protection, and rerouting.

### *ADMISSION CONTROL*

The next-hop LSR performs *admission control,* thus checking the available bandwidth for the LSP. As described in the section "Bandwidth Accounting," the available bandwidth for a certain LSP is the unreserved bandwidth plus the reserved bandwidth at all lower holding priorities. There are three possible situations.

1. If the requested bandwidth is lower than the unreserved bandwidth, it simply puts the bandwidth aside in a waiting pool and forwards the `PATH` message to the next hop specified in the ERO.
2. If the unreserved bandwidth alone is not enough but is enough together with the reserved bandwidth at lower holding priorities, one or more LSPs can be preempted. In this case, first the bandwidth is put in the waiting pool; second, the `PATH` message is forwarded to the next hop; and third, `PATHERR` and/or `RESVERR` messages are sent up- and downstream for the preempted LSPs.
3. If the available bandwidth for the LSP is not enough, a `PATHERR` message is sent back toward the head-end LSR to indicate a LSP setup failure.

### *EXPLICIT ROUTING*

As the `PATH` message is traveling across the specified path, each hop processes the ERO in the following way. First, it checks to see whether it is the next hop by comparing its own LSR ID with the LSR ID at the first position of the ERO. If yes and after performing admission control, the first position of the ERO is deleted, and the `PATH` message with the modified ERO is forwarded to the LSR denoted by the LSR ID at the next position in the ERO.

### *LABEL ALLOCATION AND LSP ESTABLISHMENT*

After the destination node has received the `PATH` message and has performed admission control successfully, it responds by sending an `RESV` message containing a `LABEL` object toward the head-end LSR following the reverse path of the ERO.

Each intermediate LSR along the path uses the label contained in a `LABEL` object of the received `RESV` message for outgoing traffic associated with the LSP. It then specifies a label and places it in the `LABEL` object of the `RESV` message to be sent to the next upstream hop. The LSR uses this specified label to identify incoming traffic associated to the traffic trunk. Both labels are used to update the LSRs *Label-Forwarding Information Base (LFIB).* In addition, the bandwidth put in the waiting pool during admission control is now really reserved when receiving the RESV message.

After the head-end LSR has received the RESV message, the LSP has been successfully set up. As a result, a forwarding state is established and represented as a route in the routing table.

### RSVP EXTENSIONS

The extensions necessary to support the required functions are specified in the IETF draft "RSVP-TE: Extensions to RSVP for LSP Tunnels" [IETF-24], as previously mentioned. The following summary is based on that IETF draft and describes the most important proposed objects used for carrying the required information, such as labels or the explicit path for an LSP tunnel.

The ERO is member of class "20," has the C_Type "1," and is used to specify the path for the LSP tunnel in the PATH message. The format of the RSVP ERO object is shown in Figure 3–52. The ERO contains a list of each hop along the path. Each hop is represented with a subobject, whereas its type denotes how the hop is characterized. Type "1" is used for IP version 4 prefixes, type "2" for IP version 6 prefixes, and type "32" for AS Numbers. The L bit of the hop subobject is set to "0" for a strict hop and to "1" for a loose hop. If the next hop is a **strict hop**, there must not be an additional node between the next hop and its preceding hop. If the next hop is a **loose hop**, there might be some nodes in between the next hop and its preceding hop.

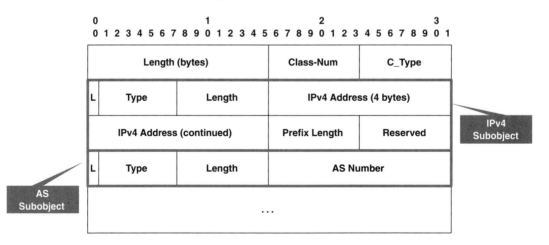

**Figure 3–52**   Format of the EXPLICT_ROUTE object

The SESSION_ATTRIBUTE object carries the information characterizing the RSVP session of the LSP tunnel. It is member of the object class "207" and has the C_Type "7." The format is shown in Figure 3–53.

```
0                   1                   2                   3
0 1 2 3 4 5 6 7 8 9 0 1 2 3 4 5 6 7 8 9 0 1 2 3 4 5 6 7 8 9 0 1
```

| Length (bytes) | | Class-Num | C_Type |
|---|---|---|---|
| Setup Prio | Holding Prio | Flags | Name Length |
| Session Name (NULL Padded Display String) | | | |

**Figure 3–53**    Format of the SESSION_ATTRIBUTE object

The 8-bit *Flag* field provides three functions: *local protection*, *LSP merging*, and *rerouting*. The flag "0x01" denotes that a downstream link or node failure can be locally protected by an intermediate node. For details, refer to the upcoming section, "Restoration," in this chapter. The Flag field set to "0x02" means that LSP merging is supported to reduce resource overhead on downstream nodes. The third flag, "0x04," indicates that the LSP may be rerouted.

The 8-bit fields *Setup Prio* and *Holding Prio* specify the value of the setup and holding priority in the range of 0 to 7.

The *Name Length* and *Session Name* fields can be used to transmit a displayable name for the LSP.

## IP Routing using MPLS-TE

Using MPLS as a forwarding engine is a simple and efficient solution for supporting explicit routing as it is required for traffic engineering. MPLS provides a complete separation of the routing and forwarding mechanism. As opposed to standard IP routing, which supports only destination-based routing, MPLS supports explicit routing. Explicit routes are calculated at the head-end LSRs, and the LFIB of all LSRs in the network is constructed in such a way that the label swapping mechanism forwards the traffic trunk's data along the calculated path.

LSPs are calculated and established, as described in the last sections. In the last step, the IS-IS or OSPF routing implementation of an LSR must be extended, because the IS-IS or OSPF routing protocol becomes aware of tunnels to make use of them.

Typically, a routing table contains numerous routes, each consisting of destination and first-hop information. The next hop is characterized using the local outgoing interface and the next-hop IP address.

To enable the routing protocol to use the LSP tunnels, the next-hop calculation must be manipulated. An LSP tunnel provides a shortcut to the tail-end LSR and can be viewed as a logical interface at the head-end LSR. As a consequence, an LSR may make use of an LSP tunnel as a next-hop interface.

The IETF draft "Calculating IGP Routes over Traffic Engineering Tunnels" [IETF-27] proposes an approach that specifies how LSP tunnels can be integrated into the IGP routing table and delivers the background for the following description.

During standard SPF computation, a router discovers the path to another router directly attached to the destination network. This router is referred to as the *next hop*. The calculating router has two possibilities when looking for the next-hop information. First, if the next-hop router is directly connected, the next-hop information is taken from the adjacency database. Second, if the next-hop router is not directly connected, it copies the next-hop information from one or more parents.

The parent of a router is the next router in the Dijkstra tree when going one step toward the root of the tree (upstream). If there are parallel equal cost paths in the Dijkstra tree, there might be multiple parents for a single router.

Within an MPLS-TE network, there must be a third possibility between the two above, considering the existence of LSP tunnels. If the next hop is not directly connected to the calculating LSR, the LSR should first consider whether there is an LSP tunnel to the next hop or to one of its upstream LSRs. If yes, use the LSP tunnel as the next hop. If not, copy the next hop from the parent(s).

Let's suppose that the network shown in Figure 3–54 is an example of how traffic is routed onto LSP tunnels. Each LSR is advertising a prefix according to its name. For example, LSR X advertises the prefix "X.X.X.X," or LSR D advertises the prefix "D.D.D.D." The routing table for LSR W is analyzed when having one LSP tunnel (T1) established to LSR D and one LSP tunnel established to LSR Z. The example also demonstrates how the traffic engineering metric influences whether an LSP tunnel or a native IP path is preferred by the IGP.

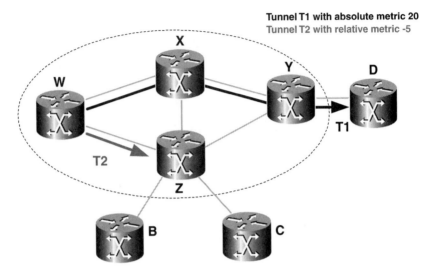

**Figure 3–54**    MPLS-TE network with two LSP tunnels

Let's first analyze the routing table with no LSP tunnels established for traffic going from LSR W to all other LSRs. The Dijkstra SPF algorithm prunes the link between LSR X and Z because the shortest path is always directly to one of the next neighbors, LSR X and Z, and is delivering the SPF tree for LSR W as output (see left half of Figure 3–55).

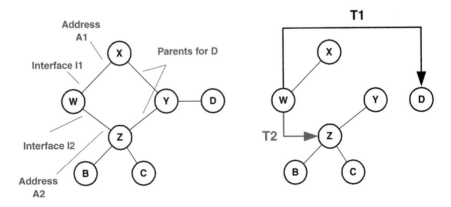

**Figure 3–55**    Shortest path, with and without LSP tunnels

In the next step, the next-hop information for each prefix is calculated. LSR X and Z are directly connected, according to the adjacency database of LSR W. The result is the routing table containing the following entries for each destination prefix in this sample network.

Prefix "X.X.X.X"—outgoing interface "I1"—next-hop address "A1."

For the prefix "Y.Y.Y.Y," LSR W is looking for LSR Y, which is not directly connected. Therefore, the next-hop information of its parent(s) is taken. Because there are two equal-cost paths, LSR Y has two parents, LSR X and Z, resulting in two entries in the routing table.

The whole routing table of LSR W is shown in Figure 3–56.

| Destination | Outgoing-Interface | Next-Hop | Metric |
|-------------|--------------------|----------|--------|
| X.X.X.X | I1 | A1 | 10 |
| Y.Y.Y.Y | I1 | A1 | 20 |
|  | I2 | A2 | 20 |
| Z.Z.Z.Z | I2 | A2 | 10 |
| B.B.B.B | I2 | A2 | 20 |
| C.C.C.C | I2 | A2 | 20 |
| D.D.D.D | I1 | A1 | 30 |
|  | I2 | A2 | 30 |

**Figure 3–56**   Routing table of LSR W with no LSP tunnels established

Now the two LSP tunnels are established. Tunnel T1 has an absolute metric of "20" assigned. Assigning an absolute TE metric means that all routes having the tunnel as next-hop have the value of the TE metric, regardless whether the routes are advertised by the tail end of the tunnel or LSRs behind. The second tunnel T2 has a relative TE of "-5" metric assigned. That means the metric used for TE depends on the IGP metrics of the traversed links.

The two tunnels are added to the SPF tree, which creates a non-tree-style graph. The SPF tree is then reestablished by first pruning the link Y-D because tunnel T1 with the absolute metric "20" is better than the path X-Y-D with a metric of "30." Second, the SPF tree is reestablished by removing link W-Z. This is because the tunnel T2 with a relative metric of "-5" is better than the link between LSR W and Z with the standard metric "10." Third, the SPF tree is reestablished by deleting link X-Y because the path to LSR Y across tunnel T2 and link Z-Y with a metric of "15" is the better choice than the native IP path X-Y with a metric of "20."

The resulting routing table is shown in Figure 3–57.

| Destination | Outgoing-Interface | Next-Hop | Metric |
|-------------|--------------------|----------|--------|
| X.X.X.X | I1 | A1 | 10 |
| Y.Y.Y.Y | T2 | Z | 15 |
| Z.Z.Z.Z | T2 | Z | 5 |
| B.B.B.B | T2 | Z | 15 |
| C.C.C.C | T2 | Z | 15 |
| D.D.D.D | T1 | D | 20 |

**Figure 3–57**    Routing table of LSR W with LSP tunnels T1 and T2 established

As can be seen in Figure 3–57, if an LSP tunnel is used as next-hop information, the outgoing interface is set to the logical tunnel interface, and the next-hop address is set to the LSR ID of the tail-end LSR.

## MPLS-TE Protection

MPLS-TE provides powerful protection functions. MPLS-TE provides flexibility when defining the network restoration strategy. It is possible to define what traffic trunks and what network resources should be protected against network failures, using the resilience trunk attribute.

Using MPLS-TE path protection, traffic trunks can be restored end-to-end by appropriately configuring the LSP tunnels used by the traffic trunks.

Using MPLS-TE link/node protection, certain network resources, such as international links or LSRs acting as gateways, can be protected by activating local protection for certain LSP tunnels.

Typically, 1:1 or 1:n protection is used within MPLS-TE. 1 + 1 protection (where traffic is sent twice over two LSPs through the network) is not used within MPLS-TE because of the enormous bandwidth inefficiency already known from SONET/SDH ring networks.

### PATH PROTECTION

When using path protection, a network failure affecting the LSP of a traffic trunk is handled by the head-end LSR through rerouting the traffic trunk onto an alternate LSP following a disjoint path. There are three options for how the head-end LSR can react to a failure situation.

For critical, high-priority traffic the resilience trunk attribute might be set to have an already precomputed and preestablished backup LSP available; thus, the switchover done by the head-end LSR can occur very quickly. To optimize the bandwidth utilization in the network, 1:n protection can be applied by specifying that multiple traffic trunks share a single LSP backup tunnel.

The resilience attribute can also specify that for important—but not very critical—traffic, the backup LSP tunnel is computed and established on demand. With this option, no idle bandwidth is reserved, but a longer restoration time must be accepted.

For best-effort Internet traffic, it might not be desired to use MPLS-TE for restoration. Therefore, the resilience attribute can specify that no backup LSP tunnel is used, and IGP convergence mechanisms are relied on to restore the failed traffic.

Figure 3–58 shows an example for option 1. According to the traffic trunk's resilience attribute, the head-end LSR is configured so that its LSP tunnel from A to B following the path A-W-Z-B is protected by the LSP tunnel following the path A-W-X-Z-B. Both LSP tunnels are up and running; in case of a failure at the link between W and Z, the head-end LSR switches the traffic onto the backup LSP tunnel and is using "9" as the outgoing label instead of "1."

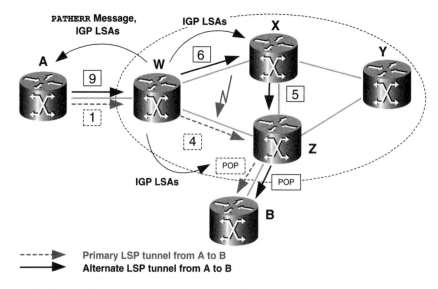

**Figure 3–58** Path protection uses backup LSP tunnel to restore the failed path

### *RESTORATION TIME*

The achievable restoration time is composed using the following delay components:

First, the failure has to be detected by the LSR closest to the failure. This is commonly done through a Hello-timer expiration at an LSR interface. Of course, this step depends on the certain layer-2 and layer-1 implementation. It might be optimized through a special failure notification function coming from layer 1/layer 2 to inform layer 3 that a failure occurred instead of waiting until the Hello-timer expires.

After detecting the failure, there are two mechanisms in the local LSR that react to the failure. On the one hand, the IGP detects the failure. After some hold-off timers have expired, LSAs are flooded, and a new SPF is calculated to adapt to the changed network topology. On the other hand, RSVP detects that the outgoing interface for the PATH state of the affected LSP failed. RSVP clears the PATH state and sends a `PATHERR` message toward the head end.

After some time also, the head-end LSR detects that there is a failure. Either its IGP process detects the failure through receiving flooded LSAs and the subsequent SPF recalculation or the LSR receives the RSVP `PATHERR` message sent out by the LSR nearest to the failure.

Now the head-end LSR can restore the traffic trunk using a backup LSP. This step can include CSPF computation and RSVP LSP setup or simply the switchover onto a preestablished backup LSP.

In summary, all these delay components lead to restoration times of several seconds.

### *OPTIMIZING PATH PROTECTION*

There are several ways for achieving a better restoration time. First, the IGP timers can be adjusted to make the IGP converging faster. By decreasing the Hello interval and the dead-timer, the failure can be recognized earlier because the Hello packets used for verifying the status of a link are sent more often. Through decreasing the SPF delay and hold time, the time between two SPF calculations is shorter. As a result, the changing topology is recognized faster. By doing this, the IGP can converge in less than 1 second.

The other big delay component is the signaling time for the alternate LSP tunnel. Through setting up multiple LSP tunnels and by using load sharing across them, no alternate LSP has to be set up in case of a failure. Thus, the signaling part can be eliminated, and only the IGP convergence at about 1 second is left.

### Link/Node Protection

In addition to path protection, MPLS also supports local protection for link or node failures. MPLS link or node protection (also referred to as *MPLS fast reroute*) provides a level of protection similar to SONET APS (SDH MSP) and achieves protection times below the 50-ms boundary.

LSP tunnels using a certain link or traversing a certain node are protected by an alternate backup LSP tunnel, which is routed around the resource to be protected. If there is a separate backup LSP tunnel for each LSP to be protected, 1:1 protection is applied. 1:n protection is used if a single alternate LSP tunnel is shared by n-protected LSP tunnels.

The setup status options for the alternate path are the same as with path protection, but only hot-standby backup LSP provides the desired protection time under 50 ms. The resilience attribute, represented by the flags of the SESSION_ATTRIBUTE object, is used to denote whether fast reroute is enabled (local protection).

In case of a failure, the affected LSP tunnels are rerouted around the failed resource across the alternate LSP tunnel. The failed LSP tunnels are nested into the alternate LSP tunnel by using the label stack mechanism shown in Figure 3–59. All actions are controlled by the upstream LSR adjacent to the failure. This LSR appends an additional label to the label stack, which then forces all LSRs to route the LSP tunnel along the path of the alternate LSP tunnel. This label is removed when the path of the alternate LSP tunnel and the path of the primary LSP tunnel merge together again.

At the same time that the local rerouting occurs, the head-end LSR of the failed LSP tunnel is notified that the LSP failed and should be reoptimized. A PATHERR message is sent toward the head-end LSR. This message contains a flag informing the head-end LSR not to clear the RSVP PATH state because MPLS fast reroute temporarily protects the failure. The head-end LSR calculates and sets up an alternate end-to-end LSP tunnel using the shared explicit mode. That means the alternate LSP tunnel does not require additional bandwidth to be reserved. The bandwidth reservation is simply transferred after the setup is completed. The traffic is switched over from the failed and temporarily protected LSP tunnel to the new end-to-end LSP tunnel.

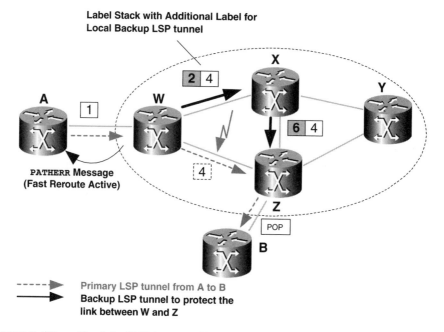

**Figure 3–59**    The failed LSP is nested into the backup LSP and routed around the failed link between W and Z

From a resilience point of view, fast reroute is optimal with already precomputed and preestablished backup LSPs. The resulting restoration time is very short. It simply depends on the time necessary for the failure detection by the upstream LSR interface. Depending on the equipment vendor's implementation, this time is typically within some ms, thus achieving the magic 50-ms boundary. The drawback is that every backup LSP consumes bandwidth, which can lead to bad bandwidth utilization in the network.

A good solution is to use fast reroute to protect multiple LSPs, which are using certain sensitive links or nodes by setting up a shared backup LSP. This might lead temporarily to local congestion. Reoptimization is provided through the use of path protection as an overlaid end-to-end mechanism.

## Network Survivability Principles

A highly available network that is very resilient to network failures is the key issue for today's network operators. As technology has evolved, the amount of traffic transported across today's high-speed networks has also increased dramatically. Even a short network outage lasting only a few seconds may cause several Gb of data to be lost if network equipment has interfaces at speeds of 2.5 Gbps or higher.

Several restoration and protection techniques have been developed over time. By applying these techniques and designing the network with enough redundant capacity, an optical network can be deployed with appropriate survivability to fulfill the requirements of all applications using the network as a central transport resource.

For more information see the whitepaper "Multi-Layer Survivability" [LUCT-1] and the descriptions of general restoration mechanisms included in the book *Optical Networks—A Practical Perspective* [RAM-1]. Using this information as a basis, the important role of the right deployment of restoration and protection techniques in a network to provide the required network resilience is to be outlined, and the design trends seen in today's networks will be reflected in the big picture.

## Defining Survivability

The term ***survivability*** is defined in [LUCT-1] as follows:

"Survivability refers to the ability of a network to maintain an acceptable level of service during a network or equipment failure. Multilayer survivability refers to the possible nesting of survivability schemes among subtending network layers, and the way in which these schemes interact with each other."

When designing a network, operators also develop a survivability concept. The included survivability strategies must be able to cope with current and future network sizes. This point is very critical because projected growths in optical networks are tremendous.

An advanced survivability concept enables the network operator to deliver a wide variety of service offerings with certain committed QoS. On the other hand, the survivability concept must guarantee the QoS commitments specified in the Service Level Agreements (SLAs).

Another important point is that some network resources may be under external administrative control, and the service offering may traverse multiple operator domains.

## Survivability Concepts

We can distinguish three basic survivability concepts (Figure 3–60).

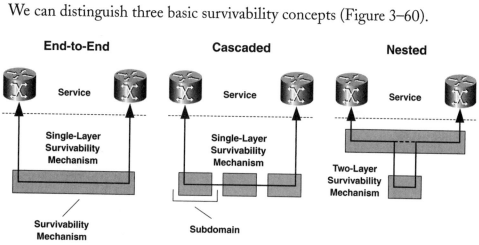

**Figure 3–60**    Survivability implementation opportunities

In *end-to-end survivability*, there is only a single survivability mechanism used to deliver end-to-end survivability.

In *cascaded survivability*, there are multiple survivability mechanisms. One mechanism is used after the other, handling faults in a certain subdomain.

The third and most common concept is *nested survivability*. Multiple survivability mechanisms are used for a single subdomain. These mechanisms may be cascaded or may also have end-to-end survivability functionality.

In general, nested survivability uses two layers of survivability, protection and restoration. This concept is shown in Figure 3–61.

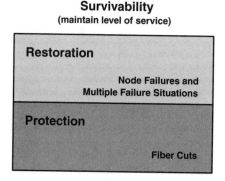

**Figure 3–61**    Nested survivability uses restoration and protection

## Protection

Protection as the lower layer mechanism provides a first level of defense against common faults, such as fiber cuts. Protection is topology- and technology-specific and offers fast recovery, but protection may be unable to protect against node failures or multiple faults. Protection is typically used in ring networks.

The protection switching intelligence is distributed onto each network element. Local defects are used as triggers. A fast detection time can be achieved because a physical media fault can be detected within several ms, for example. A fixed amount of capacity is dedicated for protection purposes to make a fast transfer of traffic from failed to good facilities possible.

Depending on how this dedicated preassigned capacity is used, we can distinguish between dedicated or shared protection mechanisms.

When *dedicated protection* is applied, 50% of the entire capacity in the network is reserved for protection purposes. It is obvious that dedicated protection delivers the highest level of protection but leads to inefficient network utilization. A typical example for dedicated protection is the Unidirectional Path Switched Ring (UPSR) architecture used in SONET/SDH ring networks.

When using *shared protection,* there is a certain amount of capacity dedicated for protection purposes and shared across the resources to be protected. A typical example for shared protection is the use of MPLS-TE. Multiple LSP tunnels traversing a common critical resource are protected by a single LSP tunnel.

## Restoration

Restoration can be seen as an overlaid mechanism, providing protection against network failures in a second step. Typically, restoration can handle not only link failures but also node or multiple concurrent failures, as opposed to protection. Restoration is typically applied in mesh topologies.

Restoration might be implemented in a centralized or distributed approach. In both cases, a network failure first must be detected locally, then must be propagated to the control element controlling the restoration procedure.

In general, distributed restoration can restore failed services faster than centralized protection. Together with the use of precomputed alternative paths, acceptable end-to-end restoration times can be achieved, which allows mesh-based, distributed restoration to become the approach of choice for implementing optical networks.

# Protection Techniques

## Protection Types

### 1 + 1 Protection

When using this type of protection, the traffic to be protected is simultaneously sent over two parallel paths. During normal operation, the destination receives two equal traffic streams and selects one of these. In case of a failure along the chosen path, the destination simply switches onto the other path.

No protection signaling is required because the destination can handle a failure by itself, and the source node does not have to do anything else but always copying traffic onto the alternate path. This makes 1 + 1 protection very simple to implement, and the achievable restoration time is very short. The disadvantage of 1 + 1 protection is the waste of bandwidth.

### 1:1 Protection

When using this type of protection, two parallel paths are also used. However, during normal operation, there is no traffic sent across the alternate path. Only in the case of a failure along the primary path do both the source and destination switch onto the protection path.

In unidirectional transmission systems, traffic is sent in only one direction over a fiber. As a result, the source does not realize a fiber cut by itself; the destination has to inform the source that a failure has occurred. In SONET networks, this signaling has been called *Automatic Protection Switching* (APS). When using bidirectional transmission systems (where traffic is transmitted in both directions over the fiber), both ends detect the failure; therefore, no signaling is required.

The obvious disadvantage of 1:1 protection is the required signaling overhead, causing slower restoration than 1 + 1 protection. More valid is its advantage. In normal operation, the unused protection path can be used for transmitting low-priority traffic, and a better network utilization can be achieved. In case of a failure of the primary path, the high-priority traffic is switched over to the protection path, and the low priority traffic is dropped.

### 1:N Protection

1:n protection is a special variation of 1:1 protection where n working paths are sharing one protection path. What is important is that this restoration technique can handle only a single failure. In case of multiple failures, the signaling protocol must ensure that only one working path is protected at a time.

### Protection Switching Characteristics

Each of these protection types can be distinguished by examining the following characteristics.

First, there are two ways of handling the traffic when a failed working path comes back again. **Nonreverting protection** does not switch back to the former working path and uses the restored path as protection path. **Reverting protection,** commonly used with 1:n protection, switches back to the former working path.

Second, who is controlling the protection procedure and how the protection path is routed through the network can be distinguished. The three possibilities were shown in Figure 3–60.

In **path switching**, which is also called **path protection**, traffic restoration is handled by the source and the destination in case of a failure, somewhere along the route between the nodes. An end-to-end restoration path is used, which is completely disjointed with the primary path.

In **line switching**, traffic restoration is handled by the nodes adjacent to the failure. Line switching can be implemented as span protection or line protection. A fiber cut between two nodes may be restored using **span protection**. By doing this, the traffic is switched onto another fiber between the same nodes. If this is not possible (because no additional fiber is available), **line protection** must be used, where the two nodes adjacent to the failed fiber are looking for a path around the failed fiber (Figure 3–62).

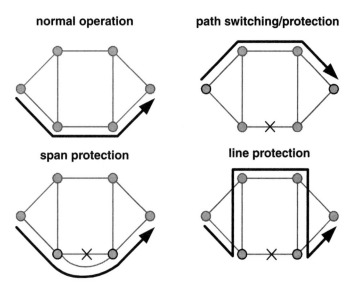

**Figure 3–62** Comparison of path and line switching after a fiber cut

## Protection in Ring Networks

Ring topologies are very popular because a ring is the simplest way to interconnect each node in a network with two other nodes to provide redundancy. Several protection switching mechanisms have been developed for SONET/SDH. These mechanisms made it possible to deploy highly available Time Division Multiplexing (TDM) ring networks with restoration times below 50 ms. The following paragraphs provide an overview about the most common SONET/SDH protection ring architectures.

However, these architectures are not solely for SONET/SDH. They have also been used for DWDM ring networks.

### Two-Fiber UPSR

As shown in Figure 3–63, UPSRs are dual-fiber rings. One ring is used as working ring, and the second ring is used for protection purposes.

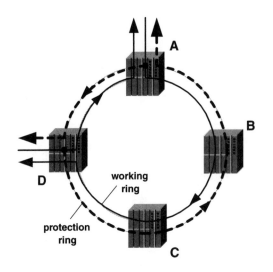

**Figure 3–63**    UPSR using 1 + 1 path protection

Traffic between two nodes is exchanged in a unidirectional manner. In Figure 3–63, traffic from node D to A is sent on the left upper part of the working ring, and the returning traffic from node A to D is sent back on the other part of the same ring. UPSRs facilitate 1 + 1 path protection. Therefore, traffic is also sent simultaneously over the protection ring in the other direction. The receiving ring node compares both signals and takes the better one.

Simplicity and a very short restoration time far lower than 50 ms are the advantages of the UPSR architecture.

An important disadvantage of the UPSR architecture is that a node failure and a failure of both the working and protection fiber of a span cannot be handled.

Another big drawback is the bandwidth inefficiency. The total ring capacity is limited to the ring speed. That means that the sum of all traffic fed into the ring must be lower or equal to the ring speed. Considering a ring at OC-48 line rate, only four OC-12 or sixteen OC-3 connections can be routed across the UPSR.

UPSRs are typically used within the metropolitan area as access rings to aggregate numerous locations into a central office. Figure 3–64 shows an example for such a "hubbed" traffic pattern. The central office is represented by node D. The locations A and C are connected each with an OC-12 pipe to the central site and the location B with two OC-12 pipes. No more locations could be added because the total capacity of the UPSR is already in use.

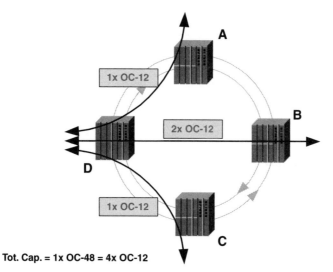

**Figure 3–64**    Total ring capacity of a four-node UPSR

The SDH counterpart of the SONET UPSR mechanism with similar functions is called *Subnetwork Connection Protection* (SNCP).

**TWO-FIBER BIDIRECTIONAL LINE-SWITCHED RINGS (2-FIBER BLSRS)**
According to Figure 3–65, 2-fiber BLSRs are also dual fiber rings. Unlike UPSRs, both rings act as working and protection rings. The bandwidth of each ring is split into two parts: One is used for carrying working traffic, and the other is used for carrying protection traffic.

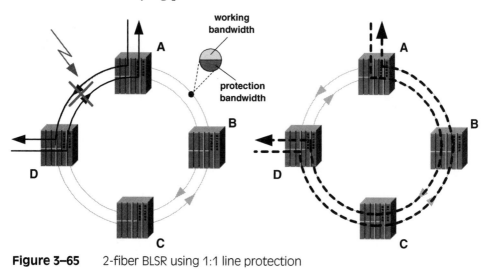

**Figure 3–65**    2-fiber BLSR using 1:1 line protection

Traffic between two nodes is exchanged in a bidirectional manner, using both rings concurrently. Looking at Figure 3–65, traffic from node D to node A is sent over the inner ring, and the returning traffic from A to D is transmitted over the outer ring.

In 2-fiber BLSRs, 1:1 line protection is used. In case of a failure, the traffic is switched onto the other fiber in the direction opposite to the failure. This failure might be a node failure or a cut of one or both fibers of a span.

There are two major advantages of 2-fiber BLSRs, compared with UPSRs. First, in case of meshed traffic patterns, bandwidth may be reused at some spans, increasing the total bandwidth of the ring. Considering again the OC-48/STM-16 ring architecture with four ring nodes, as shown in Figure 3–66, the total capacity of the 2-fiber BLSR may be up to twice the ring speed for a meshed traffic pattern. In case of the same "hubbed" traffic pattern as discussed with the UPSR, the ring capacity is quite the same as of a UPSR with four nodes. The difference is some free bandwidth on the spans A-B and B-C.

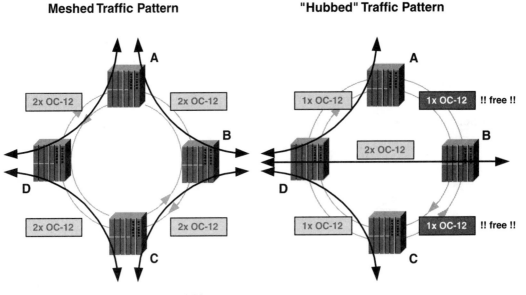

**Figure 3–66**    Total ring capacity of a four-node 2-fiber BLSR

The second advantage is that the protection bandwidth can be used to carry low-priority traffic during normal operation, because 1:1 protection is facilitated.

As a consequence, 2-fiber BLSRs are commonly used for interoffice rings, which typically have a meshed traffic pattern.

The SDH counterpart of the SONET BLSR with similar functions is called *Multiplex Section Shared Protection Ring* (MS-SPRing).

### FOUR-FIBER BIDIRECTIONAL LINE-SWITCHED RINGS (4-FIBER BLSR)

As can be seen in Figure 3–67, 4-fiber BLSRs use four fibers to interconnect the node in a ring. Two rings are used to carry working traffic; two rings are used to carry protection traffic in the opposite direction.

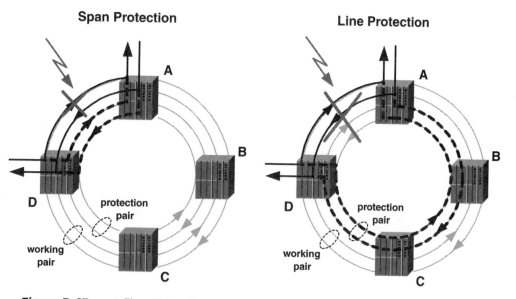

**Figure 3–67**    4-fiber BLSR using 1:1 span and line protection

Also in 4-fiber BLSRs, traffic between two nodes is exchanged in a bidirectional manner, using two rings. Looking at Figure 3–67, the four rings are separated into a working and a protection ring pair, and the two rings of the working pair are used to exchange traffic between nodes D and A.

In the case where a working fiber fails, 1:1 span protection is used. The traffic is simply switched onto the associated protection fiber on the same span. If both the working and its associated protection fiber fail, 1:1 line protection must be used. In this case, the traffic is switched onto a protection fiber in the direction opposite to the failure, as is done in 2-fiber BLSRs.

Considering again the OC-48/STM-16 ring architecture with four ring nodes, as shown in Figure 3–68, the total capacity of the 4-fiber BLSR may be up to four times the ring speed for a meshed traffic pattern. In case of the same "hubbed" traffic pattern as discussed with the UPSR, the ring capacity is quite lower because it has also been with the 2-fiber BLSR. However, in the 4-fiber BLSR, there is free bandwidth on all four spans.

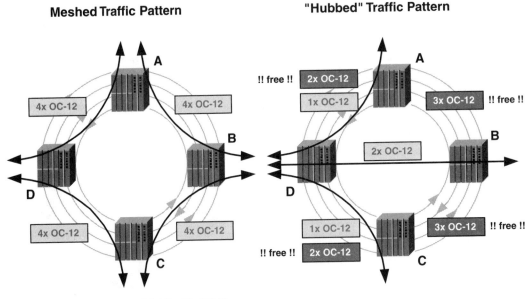

**Figure 3–68**   Total ring capacity of a four-node 4-fiber BLSR

In addition to the advantages introduced by 2-fiber BLSRs, there is the advantage of having path protection support when using the 4-fiber BLSR architecture.

Because of the bandwidth efficiency and substantial protection functionality, 4-fiber BLSRs are the preferred architecture for long-haul backbone rings.

### PROTECTION RING COMPARISON

To point out the key differences of these three ring architectures, Table 3–17 provides a short summary based on a table from the book *Optical Networks—A Practical Perspective* [RAM-1].

**Table 3–17**    Comparison of Different Types of Protection Ring Architectures

| PARAMETER | UPSR | BLSR/2 | BLSR/4 |
|---|---|---|---|
| Fiber-pairs | 1 | 1 | 2 |
| TX/RX pairs/ node | 2 | 2 | 4 |
| Spatial reuse | None | Yes | Yes |
| Protection capacity | =Working Capacity | =Working Capacity | =Working Capacity |
| Link failure | Path protection | Line protection | Span/Line protection |
| Node failure | Path protection | Line protection | Line protection |
| Restoration speed | Faster | Slower | Slower |
| Node complexity | Low | High | High |

## MPLS Restoration

The MPLS architecture defines several restoration concepts that are described in the IETF draft "Protection/Restoration of MPLS Networks" [IETF-28]. For some practical details on how protection can be implemented in data networks using MPLS-TE, refer to the previous section, "MPLS Traffic Engineering," in this chapter.

### Prenegotiated versus Dynamic Protection

When using prenegotiated protection, the backup path for each working path is configured statically. The node and link disjoint backup paths are preestablished, delivering a kind of "hot-standby" redundancy. The bandwidth of the backup paths may be predetermined and reserved in advance or they may be dynamically allocated in a failure situation by taking the necessary amount of bandwidth from low-priority traffic.

With dynamic protection, a backup path is created on demand in case of a network failure, and the traffic is rerouted onto the new alternate path. This method optimizes the bandwidth utilization in the network because no resources have to be reserved in advance.

### END-TO-END VERSUS LOCAL RESTORATION

Path protection is an example of end-to-end restoration. A failure anywhere in the network forcing the working path to fail is handled at the source of the path by setting up an end-to-end backup path. The disadvantage of this solution is the long restoration time. The failure must first be detected, then it must be propagated to the source. Finally, the traffic can be protected throughout, using the backup path.

Link protection is an example of local protection. The adjacent node detects the failure and routes the failed traffic around the failure. The protection time is much faster because there is no need to propagate the detected failure across the whole network to the source.

## Protection Switching Options

### 1 + 1 PROTECTION

When using 1 + 1 protection, a backup path is set up, and the necessary amount of bandwidth is reserved to transport an exact copy of the working traffic. At the destination, it is decided whether to take the traffic coming from the primary or backup path. Because this is not usual for MPLS, 1 + 1 is left for further study.

### 1:N AND N:M PROTECTION

A more efficient approach is to share backup paths between the working paths in the network. It is very common to use 1:n protection. One backup path is set up and used to protect the traffic if one of the n working paths fails. 1:1 protection is a special case of 1:n protection in which each working path has its own backup path.

The more complex solution also based on that sharing principle is n:m protection. The working paths are divided into groups of m paths and are protected using n backup paths.

Because n:m protection is very complex, the MPLS architecture concentrates on 1:n and 1:1 protection.

## Network Survivability Design

When designing a highly available network, many aspects must be considered. The required survivability directly affects the applications and services to be supported over the network. A key criterion is the network's restoration speed. A network carrying voice traffic must definitely recover from any network failure within a matter of ms. A network solely carrying Internet traffic may restore within only a second or less.

It is important to realize that fast restoration mostly requires a high amount of spare capacity for protection purposes. Additional wavelengths or TDM channels must be reserved in order to be available for rerouting of failed traffic.

Several restoration and protection mechanisms are available. Each mechanism has its specific advantages and disadvantages, but apart from that, the used mechanism must provide enough flexibility to adapt to changing network requirements.

### Survivability Mechanism Categories

There are numerous mechanisms used in IP, SONET/SDH, and OTNs. In general, they can be divided into three categories: dedicated protection, shared protection, and restoration.

#### DEDICATED PROTECTION
Dedicated protection is typically done at layer 1.

In SDH networks, *MSP* is used, and, in SONET networks, *APS* is used to provide a dedicated backup path in linear constellations. To provide dedicated protection in ring networks, *SNCP* is used in SDH, and *UPSRs* are used in SONET networks.

In linear constellations, providing protection at the optical multiplex section (OMS) layer might be called *Optical Multiplex Section Protection* (O-MSP) or *Optical Automatic Section Protection Switching* (O-APS). In DWDM ring networks, providing protection at the OMS layer might be called *Optical Subnetwork Connection Protection* (O-SNCP) or *Optical Unidirectional Path Switching Rings* (O-UPSR).

Because the OTN consists of three layers, protection can also be supplied at two other layers. Above the OMS layer, *optical channel protection* is used at the optical channel layer. Below the OMS layer, *line protection* is used at the optical transmission section layer.

### SHARED PROTECTION

Shared protection is also typically provided at layer 1.

In SDH networks, *MS-SPRings* and, in SONET networks, *BLSRs* are used. Part of the ring capacity (in fact, typically 50% of it) is reserved for protection purposes and is shared across all working connections in the ring.

These mechanisms can also be applied to the OTN, providing shared protection at the OMS layer. Shared protection architectures in the OTN may be called *Optical Multiplex Section Shared Protection Rings* (OMS-SPRings) or *Optical Bidirectional Line-Switched Rings* (O-BLSR).

In an MPLS IP network or IP + ATM network, shared protection can be provided through the MPLS-TE link/node protection functionality. A backup tunnel is set up around the critical link or node. The working tunnels are locally protected by switching them onto the backup tunnel in case the protected link or node fails.

### RESTORATION

Restoration is typically accomplished at either layer 2 or layer 3.

In ATM networks, the private network-to-network interface (PNNI) routing protocol is used to reroute virtual circuits (VCs) affected by a failure.

In IP networks, dynamic routing protocols, such as OSPF, IS-IS, RIP, or BGP, are used to provide rerouting in case of a failure.

A network running MPLS facilitates the possibility to define traffic trunks and to assign certain protection parameters on a per-traffic trunk basis. MPLS path protection performs rerouting onto backup paths for end-to-end service restoration.

## Restoration Time

As already mentioned, restoration speed is a key design criteria when developing a survivability concept. The ITU-T recommendation M.495 "Maintenance: International Transmission Systems" [ITU-5] specifies how the restoration time is calculated (Figure 3–69).

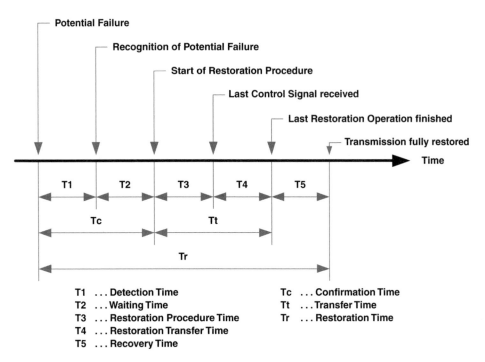

**Figure 3–69**    Restoration time components, according to ITU-T M.495

In case of a failure, it takes some time until the network node next to the failure detects the failure and triggers, for example, a Loss of Signal (LOS) or Signal Degrade (SD) event. This time interval is called ***Detection Time*** (T1).

After some time, the failure is confirmed, and the restoration procedure is initiated. The time interval between this point of time and the failure recognition is called ***Waiting Time*** (T2).

The time interval between the failure occurrence and the fault confirmation is also called ***Confirmation Time*** (Tc).

During the ***Restoration Procedure Time*** (T3), control signals are transmitted and received signals are processed.

The amount of time required for processing the last received control signal is called ***Restoration Transfer Time*** (T4).

The time interval between the fault confirmation and the point of time until the last restoration operation is finished is called the ***Transfer Time*** (T4).

In the last step, a verification of the protection switching operation or some resynchronization might be completed. This time interval is called *Recovery Time* (T5).

The overall time interval from the failure occurrence to the full restoration is called *Restoration Time* (Tr).

As an example, the detection time for SONET/SDH is specified with 10 ms and the restoration time with 60 ms. As a consequence, equipment used to deploy SONET/SDH networks must detect a failure within 10 ms and restore the failed traffic within 50 ms.

### Survivability Mechanism Comparison

The following two tables compare the most common survivability architectures regarding design criteria. These include criteria such as restoration time, prereserved restoration capacity, topology flexibility, etc.

Table 3–18 lists the architectures used in IP, ATM, and SONET/SDH networks.

**Table 3–18**    Survivability Mechanism Comparison for IP, ATM, and SONET/SDH Networks

| | MPLS RESTORATION | STANDARD IP ROUTING RESTORATION | ATM PNNI RESTORATION | SONET/SDH SHARED PROTECTION | SONET/SDH DEDICATED PROTECTION |
|---|---|---|---|---|---|
| Restoration Time | 50 ms..1 s | 1..10 s | 1..10 s | < 100 ms | < 100 ms |
| Restoration Capacity | 0..100% | ~ 0% | ~0% | < 100% | 100% |
| Rest. Cap. Usable by Low-Priority Traffic | Yes | N/A | N/A | Yes | No (UPSR, SCNP) Yes (APS, MSP) |
| Linear Topologies | Yes | Yes | Yes | Yes | Yes |

**Table 3–18**    Survivability Mechanism Comparison for IP, ATM, and SONET/SDH Networks *(cont'd)*

|  | MPLS RESTORATION | STANDARD IP ROUTING RESTORATION | ATM PNNI RESTORATION | SONET/SDH SHARED PROTECTION | SONET/SDH DEDICATED PROTECTION |
|---|---|---|---|---|---|
| Ring Topologies | Yes | Yes | Yes | Yes | Yes |
| Mesh Topologies | Yes | Yes | Yes | No | No |
| Standard-ization | In Progress | Yes | Yes | Yes | Yes |

Table 3–19 lists the architectures for the three OTN layers.

**Table 3–19**    Survivability Mechanism Comparison for Optical Transport Networks

|  | OCH PROTECTION | OMS SHARED PROTECTION | OMS DEDICATED PROTECTION | LINE PROTECTION |
|---|---|---|---|---|
| Restoration Time | < 50 ms | < 200 ms | < 200 ms | < 50 ms |
| Restoration Capacity | 100% | < 100% | 100% | 100% |
| Rest. Cap. Usable by Low-Priority Traffic | No | Yes | Yes (O-APS) No (O-UPSR) | No |
| Linear Topologies | Yes | Yes | Yes | Yes |
| Ring Topologies | Yes | Yes | Yes | N/A |
| Mesh Topologies | Possible | No | No | N/A |
| Standardization | In Progress | In Progress | In Progress | In Progress |

In general, all restoration techniques use a distributed mechanism to restore failed traffic across the whole network, in contrast to all mentioned protection mechanisms, which act locally in a point-to-point or ring subnetwork.

As a result, the restoration time achievable with one of the protection architectures is much lower and typically ranges below hundreds of milliseconds.

### Multilayer Survivability

Survivability mechanisms have to handle failures of network links, nodes, and individual channels (wavelengths). The approach followed up to now is to have a network consisting of multiple layers, to use survivability mechanisms at each layer, and to handle different types of faults in different layers.

When following this approach and designing the network for multilayer survivability, the most important parameter to focus on is the fault type and the effect that this fault has on traffic and network layers. Provisioning errors and performance degradations affect only some services at a single layer; therefore, survivability control is quite easy.

Hardware faults that force a node to fail affect all services in all layers, and have to be recovered concurrently and quickly. Handling the multiple survivability mechanisms and their reactions becomes the challenging issue.

If the detection time of the upper layer (e.g., IP) is slower than the total restoration time of the lower layer (e.g., optical layer), there will be no conflict. However, in the case where the upper layer is detecting a failure that is currently to be restored at the lower layer, interaction between the survivability processes of both layers is required to prevent the network from moving into an unstable state.

## Survivability Design Trends for Optical Networks

### Eliminating SONET/SDH and ATM

As IP traffic volumes grow, it is intended to reduce unnecessary transmission overhead. This is done through eliminating the intermediate layers ATM and SONET/SDH. The outcome is an optical network with the number of layers reduced to two.

In addition to increasing transport efficiency, this also reduces the problem of handling multiple survivability mechanisms across multiple layers, all trying to restore service simultaneously. Interworking between the optical layer and the IP layer survivability functions is not such a big issue because the optical restoration is fast enough to be transparent to the IP layer. The challenge in this new network architecture is to provide adequate survivability functions—at both the IP and optical layers.

Multiwavelength optical networks use DWDM technology to provide the required capacity. Fiber cuts are handled at the optical layer, typically using either line protection or optical channel protection. In case of a fiber cut, the IP or ATM layer does not realize any network outage because the restoration time for these two protection architectures is less than 50 ms.

If the deployed DWDM systems do not provide optical protection or if the network is simply deployed using dark fiber connections, the IP or ATM layer must also handle fiber cuts. IP or ATM routing protocols were initially not designed to restore failed traffic within such a short time. IP and ATM relied on the underlying SONET/SDH network.

As the trend goes toward optical IP networks, OSPF and IS-IS implementations are to be changed to facilitate faster convergence. By tuning keepalive intervals and hold-off timers, the IP routing protocols can converge in less than a second. Through the use of load balancing and MPLS-TE, further enhancements can be achieved. New IP optimized layer 2 technologies, such as DPT, also provide advanced survivability above layer 1. DPT's IPS functionality provides protection switching in ring networks similar to SONET/SDH.

### Putting Intelligence into the Optical Core

Wavelength routing networks can be seen as the second generation of optical networks. Network operators are facing the problem of not having a solution for getting access to the massive bandwidth provided by DWDM in an efficient and scalable way.

Typically, the OTN consists of a mesh of DWDM p-t-p systems utilizing up to 128 wavelengths. The interconnection points typically require a 3- or 4-way DWDM junction cross-connecting all these wavelengths.

The only solution for this problem is to put an intelligent device into this DWDM junction. This device, known as a ***wavelength router***, is running a dynamic routing protocol to facilitate automated wavelength provisioning. Chapter 4, "Existing and Future Optical Control Planes," describes in detail the requirements and implementation considerations for OTNs performing wavelength routing.

The first wavelength routing implementations are proprietary. They use enhanced restoration mechanisms to provide survivable end-to-end wavelength paths for interconnecting the upper-layer network equipment. These are typically IP routers. These restoration mechanisms use a distributed mechanism responsible for setting up an alternate wavelength path and rerouting the traffic onto it within less than 50 ms.

In the second stage, the standard approach of MPLmS will make it possible to deploy a peer model. As described previously, the MPLS architecture includes all necessary restoration concepts. IP routers running MPLS will have a common control plane with the wavelength routers in the OTN. They are peers from a routing perspective and directly interact in all restoration processes.

## Summary

This chapter delivered the basic knowledge about the most important optical networking technologies. We have seen that, when designing an OTN, we must consider that optical signal transmission heavily depends on the optical fiber plant. Factors such as fiber attenuation, dispersion, and nonlinear effects must be carefully watched, because they affect maximum transmission distance and maximum transmission bit rate.

WDM technology enables the transmission of multiple signals across a single fiber and supports the increase of the fiber capacity up to terabits. We covered two basic types of WDM systems: p-t-p and ring systems.

POS and DPT represent two technologies that can be used to transport IP directly over an optical infrastructure. POS is using the PPP to encapsulate IP data packets, whereas DPT is using a fundamentally new MAC layer protocol to encapsulate IP data packets. Both technologies use SONET/SDH framing and provide standard SONET/SDH physical interfaces, allowing the connection of routers as layer-3 networking devices directly to dark fiber or WDM systems.

It is obvious that network resilience becomes one of the key points in designing these next-generation OTNs. With the elimination of SONET/SDH at the lowest layer, all extensive SONET/SDH protection switching mechanisms, which provide fast state-of-the-art restoration of fiber cuts or equipment failures, are no longer available. Therefore, we need to implement equivalent protection mechanisms in OADMs to provide fast protection switching within the usual 50-ms range at the optical layer. More than that, MPLS-TE and the new state-of-the-art LSP protection mechanisms will be used to provide even higher resilience to network failures at the IP layer.

Having discussed the important optical networking technologies, we can now focus in the next chapter on suitable network architectures and control planes for next-generation OTNs.

# Recommended Reading

## Optical Transmission Technologies

[ALC-1] Alcatel Technology Paper, *Today's Optical Amplifiers—The Cornerstone of Tomorrow's Optical Layer*, Thomas Fuerst.

[IEEE-1] IEEE Communications Magazine, *Future Photonic Transport Networks Based on WDM Technologies*, Hiroshi Yoshimura, NTT Optical Network Systems Laboratories, February 1999.

[ITU-4] ITU-T Recommendation G.692, *Optical Interfaces for Multichannel Systems with Optical Amplifiers*, March 1996.

[ITU-6] ITU-T Recommendation G.652, *Characteristics of a Single-Mode Optical Fiber Cable*, 1993.

[ITU-7] ITU-T Recommendation G.653, *Characteristics of a Dispersion-Shifted Single-Mode Optical Fibre Cable*, 1993

[RAM-1] Morgan Kaufmann Publishers Inc., *Optical Networks—A Practical Perspective*, Rajiv Ramasawi, Kumar N. Sivarajan, 1998.

## Optical Transmission Systems

[CSCO-1] Cisco Systems Inc. Whitepaper, *Scaling Optical Data Networks with Wavelength Routing*, 1999.

[INFOEC-3] Technical Paper, *Position, Functions, Features and Enabling Technologies of Optical Cross-Connects in the Photonic Layer*, P. A. Perrier, Alcatel, September 1999.

## Data Transmission Technologies

### Packet over SONET/SDH

[BELL-1] Bellcore GR-253-CORE, *Synchronous Optical Network (SONET) Transport Systems: Common Generic Criteria*, Issue 2, December 1995 (Revision 1, December 1997).

[CSCO-2] Cisco Systems Inc. Whitepaper, *Cisco's Packet over SONET/SDH (POS) Technology Support*, February 1998.

[IETF-1] RFC 2615, *PPP over SONET/SDH*, June 1999.

[ITU-3] ITU-T Recommendation G.707, *Network Node Interface for the Synchronous Digital Hierarchy (SDH)*, March 1996.

[OIF-2] OIF Contribution, *A Proposal to Use POS as Physical Layer up to OC–192c, OIF99.002.2*, January 1999.

### Dynamic Packet Transport (DPT)

[CSCO-3] Cisco Systems Inc. Whitepaper, *Dynamic Packet Transport Technology and Applications Overview*, January 1999.

[CSCO-4] Networkers Conference—Session 606, *Advanced Optical Technology Concepts*, Vienna, October 1999.

[IETF-8] RFC 2892, *The SRP MAC Layer Protocol, Status Informational*, August 2000.

### MPLS Traffic Engineering (MPLS-TE)

[IETF-17] IETF-draft, draft-katz-yeung-ospf-traffic-01.txt, *Traffic Engineering Extensions to OSPF*, work in progress, October 1999.

[IETF-23] IETF-draft, draft-ietf-isis-traffic-01.txt, *IS–IS Extensions for Traffic Engineering*, work in progress, May 1999

[IETF-24] IETF-draft, draft-ietf-mpls-rsvp-lsp-tunnel-05.txt, *RSVP-TE: Extensions to RSVP for LSP Tunnels*, work in progress, February 2000.

[IETF-25] IETF-draft, draft-ietf-mpls-cr-ldp-03.txt, *Constraint-Based LSP Setup Using LDP*, work in progress, September 1999.

[IETF-26] RFC2370, *The OSPF Opaque LSA Option*, July 1998.

[IETF-27] IETF-draft, draft-hsmit-mpls-igp-spf-00.txt, *Calculating IGP Routes over Traffic Engineering Tunnels*, work in progress, December 1999.

[JUNP-1] Juniper Networks Whitepaper, *Traffic Engineering for the New Public Network*, January 1999.

[JUNP-2] Juniper Networks Whitepaper, *RSVP Signalling Extensions for MPLS Traffic Engineering*, August 1999.

### Network Survivability Principles

[IETF-28] IETF-draft, draft-makam-mpls-protection-00.txt, *Protection/ Restoration of MPLS Networks*, work in progress, October 1999.

[LUCT-1] Lucent Technologies Whitepaper, *Multi-Layer Survivability*, J. Meijen, E. Varma, R. Wu, Y. Wang, 1999.

[RAM-1] Morgan Kaufmann Publishers Inc., *Optical Networks—A Practical Perspective*, Rajiv Ramasawi, Kumar N. Sivarajan, 1998.

# 4

# Existing and Future Optical Control Planes

**O**ne of the main goals of this book is to point out how optical network architectures are changing to adapt to new traffic patterns and the introduction of new types of Internet Protocol (IP)-based services.

Traditional old-world multilayer networks have been deployed using different layer-1 and layer-2 technologies, such as Synchronous Optical Network (SONET)/Synchronous Digital Hierarchy (SDH) and Asynchronous Transfer Mode (ATM), as a transport infrastructure for IP. The enhancement of IP routers to provide the routing performances up to the gigabit (or, in the future, even terabit) levels, as well as the commercial availability of high-bandwidth router interfaces such as 2.5 Gbps or 10 Gbps, marked the first step in the evolution of IP service-based carrier networks.

In this first step, the IP network resides directly on top of a static Optical Transport Network (OTN), and the control between IP and optical is provisioned in a static way. Thus, it can be also referred to as a network with a **_Static IP Optical Overlay Control Plane_**.

Further on, the OTN is transitioning from a ring or point-to-point (p-t-p)-based network into a network using a mesh topology with multiple Dense Wave Division Multiplexing (DWDM) junction points; the old control plane of static provisioning is no longer suitable. This evolution marks the second step, where dynamic wavelength provisioning is introduced to allow easier, faster, and more scalable control of the OTN. It will also deliver enhanced restoration capabilities. Because the connectivity through the OTN is now delivered in a dynamic fashion, we speak of this advanced network design using the term ***Dynamic IP Optical Overlay Control Plane***.

Because carriers have both the OTN and the IP network under their control, they are looking for further ways in which to streamline their network. The increasing dynamic nature of today's and future IP services is creating frequent requirements for changing connections, primarily in terms of bandwidth on the OTN. Thus, efforts are being undertaken to integrate the dynamic wavelength provisioning processes of the OTN into the IP network routing mechanisms, making the OTN visible to the IP network, and to create an ***Integrated IP Optical Peer Control Plane***.

The Static IP as well as the Dynamic IP Optical Overlay Control Plane have several things in common. The OTN is opaque to the IP network and is simply providing connections to the above IP network. Furthermore, the route finding of these connections is not controlled by an IP-aware routing intelligence. These architectures are commonly referred to as the ***overlay model***—an example being the IP/ATM overlay model (see also "IP and ATM Overlay," described in [IETF-1]).

This type of architecture is fundamentally different from the ***peer model***, which is using an Integrated IP Optical Peer Control Plane. In the peer model, a common control plane for both the OTN and IP networks is deployed, and the optical connections are derived from IP routing knowledge. The concept of both models is illustrated in Figure 4–1.

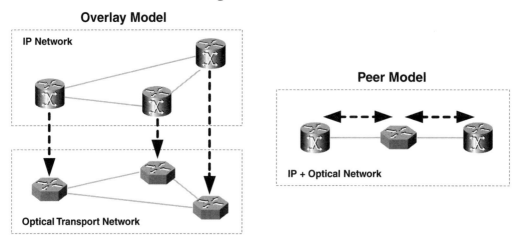

**Figure 4–1**    In the overlay model, the OTN is opaque to the IP network, whereas, in the peer model, the OTN is integrated into the IP network, thus becomes visible to it

We will describe these three methods of controlling an OTN throughout this chapter in detail.

# Static IP Optical Overlay Control Plane

The elimination of intermediate layers leads to a fundamentally new approach of building IP networks. The result is a new architectural model with tremendous efficiency and scalability required for coping with upcoming network requirements.

The topology of the service infrastructure is one key point in the design of optical networks. Furthermore, the effects of introducing multiwavelength transmission through the use of DWDM technology have to be taken into account when implementing IP-centric service infrastructures.

## Static IP Optical Overlay Model

### IP Service Infrastructure

Network services are increasingly moving toward IP. Most of today's applications are based on the TCP/IP Protocol. In the future, all applications will be. Today's approach to transport IP traffic over Time Division Multiplexing (TDM) networks formerly optimized for transport of isochronous traffic like voice becomes inefficient and obsolete.

Traditional networks have used SDH for delivering reliable Wide Area Network (WAN) connectivity at layer 1. Those SDH networks have consisted of optical rings connecting SDH add/drop multiplexers (ADMs) and digital cross-connects (DXCs) for interconnecting the SDH rings. Because of the complex multiplexing structure of SDH, a massive amount of equipment has been required.

ATM has been commonly used for the delivery of data services above the SDH network. ATM with its connection-oriented approach enables the network operator to deliver connectivity to customers with certain Quality of Service (QoS) requirements and to provide Virtual Private Network (VPN) services.

Networks based on TDM and ATM technology (also commonly referred as *"old-world" networks*) produce an excessive amount of operational costs and facilitate very low transport efficiency. When transporting packet-oriented protocols such as IP over an ATM infrastructure, a huge overhead is introduced through segmentation and reassembly (SAR) functions. The typical ATM framing overhead is about 20%. As discussed in "Traffic Engineering for the New Public Network" [JUNP-1] the overhead introduced by ATM on an OC-48/STM-16 link results in only about 2 Gbps bandwidth available for transmitting data traffic. This means the capacity of nearly a whole OC-12/STM-4 link is wasted. Going a step further and looking at an OC-192/STM-64 link, nearly the capacity of an OC-48/STM-16 link is wasted. As technology evolves and the industry constantly demands interfaces with higher and higher bit rates, ATM's inefficiency is not acceptable any longer, and router networks using alternative technologies such as Packet over SONET/SDH (POS) and Gigabit Ethernet are to be favored.

For detailed overhead calculations, refer to the section "Technology Comparison" in Chapter 3, "Optical Networking Technology Fundamentals."

Implementing an optical network, thus eliminating ATM and SONET/SDH equipment, delivers a tremendously simplified two-layer network often referred to as a *"new-world" network*. It consists of an IP service infrastructure and an optical transport infrastructure, as shown in Figure 4–2.

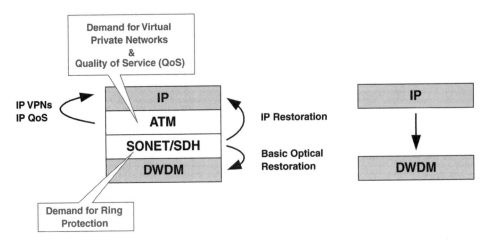

**Figure 4–2**   Advanced functionality in the IP and optical layer

The optical transport infrastructure is built using dark fiber connections and DWDM equipment. The optical transport infrastructure using DWDM technology is often referred to as the *optical layer* and provides optical paths (wavelengths), commonly known as *optical channels* and used most commonly for transporting IP but also for other types of data or voice traffic.

### IP/Optical Adaptation

One major advantage of IP is that it can be transported over most data link protocols and underlying networking technologies. This is a result of decoupling IP from the transport mechanisms.

To transport IP over a wavelength, IP traffic must be encapsulated, and the encapsulated traffic must be inserted into a defined optical interface framing. The National Fiber Optic Engineers Conference (NFOEC) technical paper " 'IP over WDM' the Missing Link" [NFOEC-1] summarizes the methods that have been developed in recent years. These are shown in Figure 4–3.

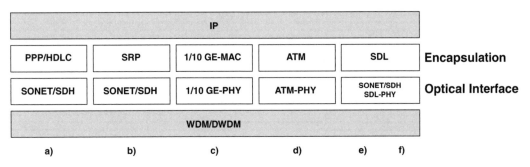

**Figure 4–3**   Protocol stacks for mapping IP onto DWDM

Each of these protocol stacks can be differentiated in several ways. First, how much bandwidth overhead is introduced? Second, what interface line rates are available? Furthermore, how scalable is the network solution, based on the certain technology? What traffic management functions are available? Also, of course, what are the QoS capabilities? These are very important when building networks capable of supporting multimedia applications and real-time traffic.

The listed technologies are currently available, but it is very likely that there will be several new protocols for high-bandwidth networks.

- *POS (a):* The first protocol stack, POS, uses Point-to-Point Protocol (PPP) for encapsulation and the SONET/SDH physical interface to adapt to the optical channel. POS has already been covered in detail in Chapter 3, "Optical Networking Technology Fundamentals."

- *Dynamic Packet Transport (b):* Dynamic Packet Transport (DPT) uses a MAC layer protocol called **Spatial Reuse Protocol** (SRP) for IP packet encapsulation. It also uses SONET/SDH as a physical interface to DWDM. DPT has been covered in detail in Chapter 3, "Optical Networking Technology Fundamentals."
- *Gigabit Ethernet (c):* Extensive developments in the Gigabit Ethernet area brought up very long-reach interfaces and DWDM equipment-compatible optical interfaces. Thus, the third approach is to use the well-known Ethernet MAC layer together with the new 1-Gbps and 10-Gbps physical interfaces to transport IP traffic over optical paths. Especially for networks in the metropolitan area, Gigabit Ethernet can be used to seamlessly interconnect multiple campus networks over dark fiber or a DWDM transport network.
- *ATM (d):* The fourth possibility is the elimination of the intermediate SONET/SDH layer and the direct mapping of ATM cells into WDM channels using optical high-speed ATM interfaces.
- *Simple Data Link (e, f):* Simple Data Link (SDL) also introduces a new data link protocol that can be used with either SONET/SDH physical interfaces or SDL native physical interfaces.

### Two-Layer Architecture Implementation

Figure 4–4 shows the new-world architecture, which consists simply of two layers in detail. IP routers at the edge of the network aggregate the ingress traffic delivered by several access technologies, such as ADSL, cable, dial, or leased line services and statistically multiplex that traffic onto *"big fat pipes"* (BFPs).

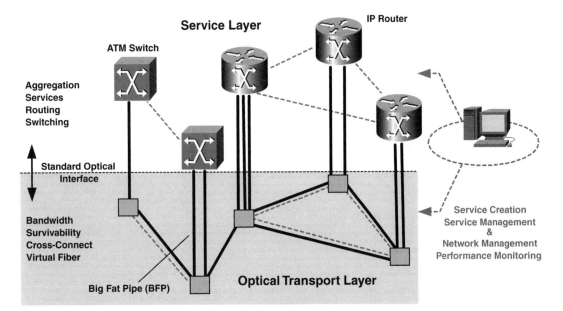

**Figure 4–4**   The New-World Architecture and its functions incorporated by the service layer and optical transport infrastructure

These BFPs delivering "raw bandwidth" are provided by the optical layer, through either dark fiber connections or DWDM wavelength channels between routers. The optical layer can be seen as a cloud for interconnecting attached devices, such as SONET/SDH nodes, ATM switches, or routers of the service layer. The connections can be either p-t-p, ring, partially meshed, or fully meshed, depending on how the DWDM optical backbone is configured. This is analogous to running IP networks over an ATM infrastructure, meaning that any logical connectivity can be designed between IP nodes via configuring the appropriate VCC/VPC connections. Currently, when using dark fiber, connections up to 10 Gbps can be delivered using either SONET/SDH or Gigabit Ethernet-based interfaces. This will probably change as technology evolves; 40-Gbps transmission has already been tested in field trials. When using DWDM p-t-p systems, much higher transport capacity can be achieved. Multiple wavelengths are transmitted, each carrying a high bit rate signal, providing bandwidths up to the terabit level. Multiple router interfaces are used in parallel for one connection and are connected to a DWDM terminal.

DWDM ring systems are very common, especially for metropolitan area applications. Not only does the bandwidth multiplication principle described before apply to DWDM rings, but also enhanced optical restoration functions can be provided. Ring protection mechanisms well known from SONET/SDH rings are also implemented in optical rings for delivering high optical layer resilience.

Optical layer restoration is covered in detail in the section "Restoration in Static Optical Networks," later in this chapter.

Implementing optical networks facilitates high-bandwidth connections via optical path provisioning with core scalability independent from the IP, ATM, or SONET/SDH aggregation. Core routers directly interface to the optical layer; thus, no inefficiency through intermediate multiplexing schemes, as with SONET/SDH networks, occurs.

IP aggregation rings, implemented through DPT, utilize not only statistical multiplexing, but also multicasting, fast restoration, and maximum bandwidth efficiency through spatial reuse capabilities. This is very important for sites mainly delivering IP and, particularly, those delivering multimedia application services. For a detailed description of DPT, refer to Chapter 3, "Optical Networking Technology Fundamentals."

Another important part of the new-world architecture is a common management system. If all devices are controlled within a single management domain, operational costs can be dramatically decreased, and management processes can be simplified. This unified management plane handles device configuration, service provisioning, performance monitoring, and all other management activities. There might be some proprietary management systems for vendor-specific element management functions, but a standard interface should ensure information exchange with the central management system.

## Multiwavelength Transmission

Dark-fiber connections and DWDM systems are used to build an optical transport infrastructure. This infrastructure provides statically configured optical paths. Those optical paths are used for interconnecting routers, ATM switches, and SONET/SDH equipment.

DWDM technology allows multiple optical connections to be run simultaneously over dark-fiber and thus makes bandwidth scalable. Further bandwidth can be easily added by assigning additional wavelength channels. A typical topology consisting of a mix of ring and p-t-p interconnections is shown in Figure 4–5.

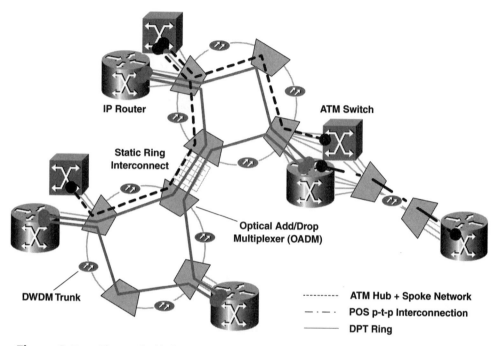

**Figure 4–5**   The optical infrastructure provides the logical connections through static wavelength provisioning

This architecture is far away from the "ideal" optical network, which can provide automated end-to-end wavelength services. Within this architecture, all networking features are implemented into the IP service layer. The optical layer simply provides "big dumb pipes" connecting each service layer device with the next hop.

A further drawback in this stage of optical internetworking is that DWDM creates multiple overlaid networks. Multiple virtual p-t-p, ring, hub + spoke, or meshed style topologies are set up through statically assigning wavelength channels. These networks provide connectivity between the routers attached to the optical network.

DWDM rings provide any virtual topology, but this flexibility is limited to each ring. Optical DWDM rings consist of OADMs interconnected via DWDM trunk ports. The OADMs are providing multiple local interfaces for connecting routers with high-speed data interfaces. As shown in Figure 4–6, an optical ring can provide the optical paths required for establishing the desired logical topology. This might, for example, be a logical mesh of p-t-p connections using POS, or a group of p-t-p connections forming a logical ring while using DPT.

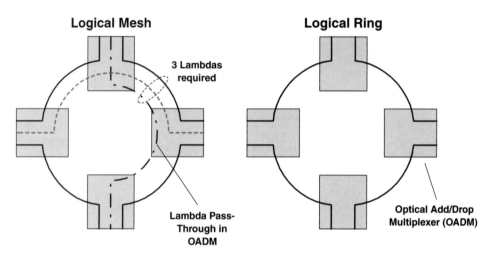

**Logical Mesh**

3 Lambdas required

Lambda Pass-Through in OADM

**Logical Ring**

Optical Add/Drop Multiplexer (OADM)

**Figure 4–6**    Optical paths for establishing required logical topologies

Ring and optical channel interconnection can be implemented within the optical domain only via static optical cross-connects (fiber patch-panels). This static approach leads to excessive operational costs and requires manual reconfiguration in case of network failures.

An alternative is to terminate the optical channels at the DWDM ADMs or p-t-p terminals and to use routers or ATM switches for interconnecting the number of optical channels at the service layer. In that way, the number of multiwavelength trunks to be interconnected is limited to the interface density and forwarding capacity of the used routers and ATM switches.

End-to-end service provisioning and network restoration is the major problem. Optical restoration is limited to a single "*optical subnetwork*," represented by either a DWDM ring or a p-t-p system. As a consequence, bandwidth management and survivability functions are implemented in the IP service layer.

Survivability issues are discussed in detail in the section "Restoration in Static Optical Networks" of this chapter.

## Bandwidth Management

Functionalities such as QoS, restoration, and VPNs, formerly provided by ATM and SONET/SDH, must now be provided through the remaining layers. The demand for virtual networks with specific QoS, covered by ATM with its virtual circuits (VCs) and virtual paths (VPs) laid over the whole network in a meshed style, is now done by IP QoS, Multiprotocol Label Switching Traffic Engineering (MPLS-TE), and MPLS VPNs.

To ensure a proper transmission of traffic within specific constraints defined by parameters such as delay or jitter, IP QoS is used. IP QoS can basically be divided into the following parts, as listed in [CSCO-9]:

- *Classification* entails using a traffic descriptor to categorize a packet within a specific group, to define that packet, and to make it accessible for QoS handling on the network.
- *Congestion management* entails the creation of queues, assignment of packets to those queues based on the packet's classification, and scheduling of the packets in a queue for transmission.
- *Congestion avoidance techniques* monitor network traffic loads in an effort to anticipate and avoid congestion at common network bottlenecks. Congestion avoidance is achieved through packet dropping.
- Both *policing and shaping mechanisms* use the traffic descriptor of a packet—indicated by the packet's classification—to ensure adherence and service.
- *QoS signaling* is a form of network communication that provides a way for an end station or network node to communicate with its neighbors to request special handling of certain traffic.

Thus, traffic is assigned to certain traffic classes with specific Service Level Agreements (SLAs). To obey the SLAs, queuing and congestion avoidance mechanisms are used throughout the whole network.

The simplest way to access the massive bandwidth provided by DWDM is to use load balancing. If there are two or more optical paths between two routers, there are multiple routes in the routing table. If the optical paths have the same line rate, the routing protocol automatically spreads the traffic onto the several parallel optical paths. In case of unequal optical path speeds, the routing protocol must be tuned via metric manipulation to ensure multiple routes with equal metrics and, therefore, provide load balancing.

The more advanced method to access the bandwidth is to use MPLS-TE, which gives the network operator a very powerful tool to control the traffic flow in its network. Tunnels can either be set up statically by using an off-line management tool or dynamically by using Resource Reservation Protocol/Label Distribution Protocol (RSVP/LDP) signaling to ensure optimum network utilization according to the actual traffic situation.

MPLS-TE is based on MPLS, which is a new forwarding technique integrating layer 2 and layer 3. MPLS can be deployed in either router-only or IP + ATM networks. MPLS uses fixed-length labels appended to the IP packet header or inserted in the ATM cell header. These labels are inserted/removed at the edge of the MPLS network by edge label switch routers (Edge-LSRs) and are used by label switch routers (LSRs) for taking the forwarding decisions in the network.

MPLS-TE makes use of MPLS and establishes traffic trunks with certain attributes to satisfy the required QoS and restoration level. These traffic trunks can be established either statically or dynamically and can be restored at certain levels for delivering the necessary network resilience. MPLS-TE is discussed in depth in Chapter 3, "Optical Networking Technology Fundamentals."

Both load balancing and traffic engineering are also covered a bit more in detail in the section "Restoration in Static Optical Networks" of this chapter when analyzing restoration methods used in multiwavelength optical networks.

## Static Optical Control Plane

Considering the static IP optical control plane in the overlay model, the IP service infrastructure based on gigabit routers equipped with high bit-rate interfaces is statically connected across an optical infrastructure based on DWDM or dark fiber.

### IP Service Infrastructure Implementation

How the gigabit routers of the IP service layer are interconnected from a topology point of view and how they are used together with DWDM equipment is described in the following sections.

Note that the following topology considerations for the IP service infrastructure do not take into account geographical restrictions specific to each individual network implementation. Keep in mind that this does not mean that geographic restrictions are not important. They have a significant impact on how the backbone is structured but cannot be covered in theory. Available fiber connections and distances between locations are different in every real-world application.

**SMALL POINT-TO-POINT BACKBONES OR BACKBONE INTERCONNECTIONS**
Small backbones or long-haul backbone interconnections are built with p-t-p fiber connections. To provide protection in case of fiber cuts, signal degrade, or interface failures, each backbone router is equipped with two POS high-speed interfaces and connected to the neighbor router over two fibers, as shown in Figure 4–7.

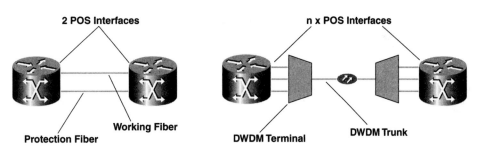

**Figure 4–7**    Optical p-t-p backbone interconnection with dark-fiber or DWDM terminals

Through the use of two interfaces, load balancing can be applied. If one link fails, the traffic is carried by the remaining link, and service interruption can be minimized. For details on network restoration, refer to a following section, "Restoration in Optical Networks."

Higher bandwidth interconnections can be implemented by using DWDM systems, as shown in the right half of Figure 4–7. Multiple POS interfaces are directly connected to the DWDM node, which dedicates a wavelength to each POS interface and multiplexes them onto the fiber. Redundancy is added by using two fibers between the DWDM nodes. The advantages of using DWDM technology are the higher bandwidth over one fiber and the fiber protection supported by the DWDM system, which is transparent to the attached routers.

### Large-Scale Hierarchical Backbones

For backbones consisting of more than two nodes, a meshed topology, as shown in Figure 4–8, can be considered. To create a fully meshed topology,

$$\frac{n*(n-1)}{2}$$

links have to be established, and n-1 interfaces on each router are required. The advantage is that every node is reachable via one hop; thus, long multihop paths are avoided. The disadvantage is the tremendous amount of required connections and interfaces.

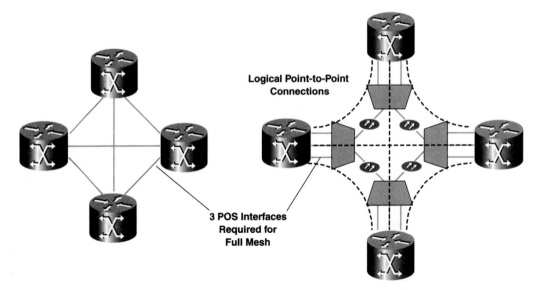

**Figure 4–8**  Fully meshed POS backbones can be implemented either through using dark-fiber connections or DWDM systems

Using DWDM technology again, the bandwidth between backbone routers can be increased without additional fibers, just by adding additional interfaces. Also, protection mechanisms are transferred to the DWDM nodes. The DWDM nodes are commonly connected to a fiber ring, eliminating the need for

$$\frac{n*(n-1)}{2}$$

fiber connections. These nodes can provide any topology to the routers: ring, meshed, or hub, as described in a previous section, "Two-Layer Architecture Implementation."

For large service provider backbones with about 10–20 core nodes and about 50–100 Access Points of Presence (Access-POPs), the meshed approach does not scale well, because interconnecting 15 core routers fully meshed, for example, would require 105 fiber connections. Networks of that size have to be hierarchical structured. A limited number of nodes are placed in the core. A ***distribution network*** is established around the core to aggregate the high amount of POPs.

The topology of the core can be either mesh or ring style. To work out the differences of both possibilities, a network with eight nodes in the core is assumed. Connecting them fully meshed would require 28 fiber connections for the core. This is a requirement that can hardly be fulfilled. However, when putting three nodes into a triangle and dual-homing the remaining nodes around that triangle, the required fiber connections are reduced to 13. Figure 4–9 shows this solution. If any fiber connection or node fails, there is always another path to restore the desired path.

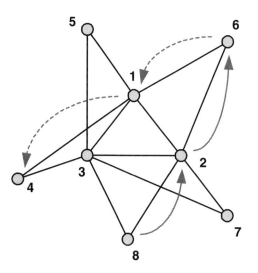

**Figure 4–9**    Connecting the core nodes in a mesh requires a lot of fiber connections

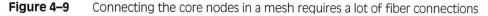

The other possibility, shown in Figure 4–10, is connecting the nodes in ring style with POS p-t-p connections. With eight core nodes, two rings can be formed by interconnecting via two nodes. This is important to provide redundancy and prevent splitting up the network into two parts, in case the interconnecting node fails. This solution requires nine fiber connections.

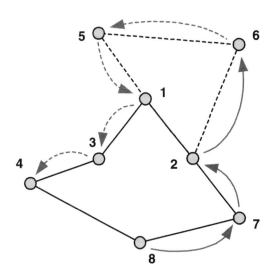

**Figure 4–10**   Connecting the core nodes in one or more rings reduces the amount of required fiber connections

The advantage of the meshed topology is that there is always a path available to transit the core with only one intermediate router hop, regardless of where the core is entered and left. With the ring solution shown in Figure 4–10, it can take up to three intermediate router hops to transit the core. The advantage of the ring topology is the small amount of fiber connections needed. Assuming costs and availability of dark fiber, the ring topology is the solution of choice.

DPT technology supports the implementation of high-performance ring-based architectures as shown in Figure 4-11. Existing dual-fiber rings formerly used for SONET/SDH services can be reused. Besides the fact that DPT allows packet-oriented communication across the ring, the limit of a maximum of 16 SONET/SDH nodes per ring is no longer valid, allowing up to 256 nodes (theoretically).

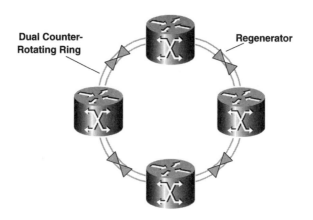

**Figure 4–11**   Optical backbone ring using SRP

Only one interface is required in each backbone router, independent from the number of routers. The interface provides two ports: one for the east and one for the west neighbor. This is a significant advantage, compared with optical rings using POS technology, where two interfaces per node are required. Furthermore, transit traffic can go through the transit buffers of the DPT interfaces and does not have to be processed by the layer-3 routing engine of router. For neighbor distances longer than 40 km, IP regenerators are used to ensure proper data transmission.

As discussed in the whitepaper, "Dynamic Packet Transport Technology and Applications Overview" [CSCO-3], DPT rings address all challenges of IP transport rings. DPT can handle the growing IP traffic volume and the massive fiber bandwidth growth through DWDM. The SRP fairness algorithm (SRP-fa) maintains high-bandwidth utilization and high throughput, even under congestion situations, and can rapidly adapt to changing traffic patterns. DPT can be used for large-scale ring topologies, both in terms of the number of nodes, as well as distances between the nodes. It is a solution for metropolitan and wide area rings as well as intrabuilding rings. Enhanced queuing and priority mechanisms support multiple traffic classes with varying bandwidth and delay requirements. Intelligent Protection Switching (IPS) enables rapid recovery from faults, including transport media and node failures. A dynamic topology recovery function allows plug-and-play operation for insertion, removal, and recovery of ring nodes with minimal configuration and provisioning requirements.

Optical fiber rings using DPT technology can be set up above dedicated fiber connections or DWDM systems, as well as hybrid infrastructures. Long distances between the ring nodes can be achieved, especially with DWDM systems. This can lead to big transmission delays, which have to be considered to avoid timing problems within the SRP-fa signaling mechanism of DPT.

DPT technology was discussed in depth in Chapter 3, "Optical Networking Technology Fundamentals."

### POP AGGREGATION

POPs provide the access for customers to the service provider network. Depending on the required services, the POPs consist of routers, ATM switches, xDSL concentrators, or cable head ends and can be connected to the backbone routers in several ways. The basis for the following descriptions has been provided by the whitepaper "Building High-Speed Exchange Points" [CSCO-8], which contains detailed information on POP implementations.

In a second stage, these numerous Access-POPs are aggregated in POPs of the distribution layer of the network. The Distribution-POPs should be designed to leverage triple-layer POP redundancy. First, there should be two backbone edge routers load balancing the traffic into the backbone. For situations in which one of these routers fails, it is important that the other router is able to take the full load by itself. Second, both of these routers should have two connections to the backbone. Third, there should be two POP interconnect devices and/or a physical fail-over medium.

### *SWITCHED POP AGGREGATION*

One solution can be to use layer-2 switches between the Access-POPs and the backbone edge routers (Figure 4–12). ATM or Ethernet can be considered. The drawback of ATM is the high equipment cost. This cannot be compensated with the powerful QoS features of ATM because there is enough bandwidth available, anyway.

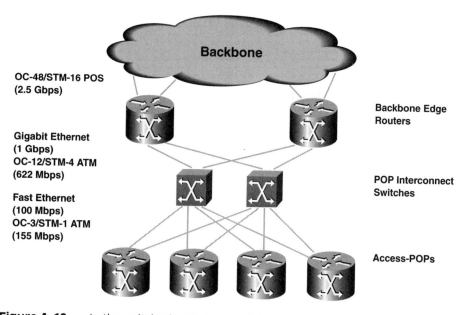

**Figure 4–12**    In the switched POP design, all Access-POPs and backbone edge routers have a connection to both layer-2 switches

When using Ethernet technology, Fast-Ethernet and Gigabit-Ethernet interfaces are appropriate today, and 10-Gigabit Ethernet coming soon. Between the Access-POPs and the backbone edge routers, two switches are placed to provide redundancy. Typically for today, each Access-POP is connected to both switches with Fast-Ethernet uplinks. The backbone edge routers are connected with gigabit uplinks to the switches. Thus, a hierarchy in bandwidth with 100 Mbps from the Access-POP, 1Gbps to the backbone edge router, and 2.5 Gbps into the core backbones is achieved.

To prevent bridging loops caused by the dual-homed routers, the Spanning Tree algorithm (STA) has to be deployed.

There are multiple advantages to this approach. First of all, Ethernet is an easy-to-deploy and proven technology. An Ethernet solution is scalable to adapt to future bandwidth requirements, as the evolution from 10 Mbps to 100 Mbps and 1,000 Mbps Ethernet has shown. Furthermore, scalability is provided through modular switches, which can easily be equipped with more uplinks in the layer-2 switch for connecting additional routers. The disadvantage is that there are additional switches providing additional sources of failures.

When using ATM today, available interface rates are OC-3c/STM-1 at 155 Mpbs, OC-12c/STM-4c at 622 Mbps, or OC-48c/STM-16c at 2.5 Gbps. Similar to the Ethernet solution, two ATM switches are put between the Access-POP and the backbone edge routers to provide redundancy. The Access-POPs may, for example, be connected with OC-3c/STM-1 uplinks to both switches, and the backbone edge routers may be connected with OC-12c/STM-4c uplinks.

To integrate the connectionless IP world into the connection-oriented ATM world, further processes have to be deployed. The first solution is Classic IP (CLIP), where IP routes are mapped into ATM VPs. With the second solution, called *LAN emulation* (LANE), a LAN would be established virtually between the access and backbone routers. The specific abilities of these technologies cannot be discussed in this book. However, both result in additional overhead and implementation complexity. For further information on IP over ATM, refer to [AF-1], [CSCO-7], [MINO-1], [IETF-6], [IETF-7], or [CSCO-6].

One advantage of ATM-switched POP aggregation is that ATM facilitates a fully meshed POP interconnection. Thus, direct connections between the Access-POPs for inter-POP traffic are possible. The main disadvantages are the high equipment costs of ATM and the tremendous overhead required to integrate IP and ATM.

### POINT-TO-POINT POP AGGREGATION

POS, ATM, or Gigabit-Ethernet may be used for deploying the p-t-p connections. Each Access-POP is directly connected p-t-p to both backbone edge routers.

Today's POS or ATM interconnections may be made of OC-3c/STM-1 links at 155 Mbps or OC-12c/STM-4c links at 622 Mbps. With OC-192c/STM-64c interfaces in the backbone in the near future, OC-48c/STM-16c interfaces at 2.5 Gbps will also be used for POP aggregation.

The main advantage of this solution is that there is no interconnect device necessary. When using POS, there is the same technology used both in the backbone and in the distribution layer (Figure 4–13).

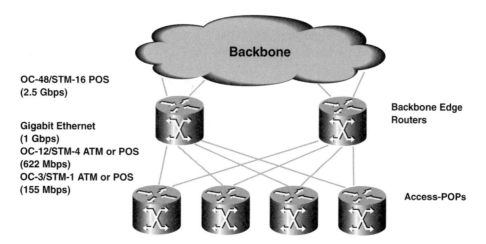

**OC-48/STM-16 POS**
**(2.5 Gbps)**

**Gigabit Ethernet**
**(1 Gbps)**
**OC-12/STM-4 ATM or POS**
**(622 Mbps)**
**OC-3/STM-1 ATM or POS**
**(155 Mbps)**

**Figure 4–13**    Interconnecting the Access-POPs directly with the backbone edge routers eliminates the need for an interconnect device

The main disadvantage is that there is an interface for each Access-POP required in the backbone edge router, which, as a consequence, must have a very high interface density.

### DPT FIBER RING POP AGGREGATION

When using DPT technology, an access ring is established, containing both the backbone edge routers and Access-POPs (Figure 4–14). This replaces the dual-homed hub topologies. The bandwidth of the access ring depends on the DPT interfaces used. Currently, there are OC-2c/STM-4c or OC-48c/STM-16c interfaces available. In the near future, OC-192c/STM-64c interfaces will also be implemented.

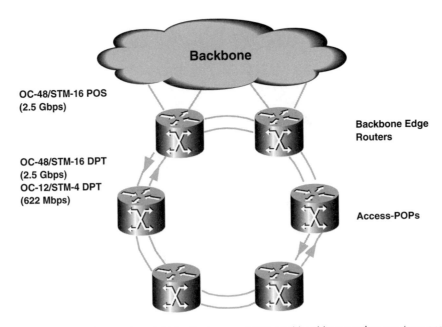

OC-48/STM-16 POS
(2.5 Gbps)

OC-48/STM-16 DPT
(2.5 Gbps)
OC-12/STM-4 DPT
(622 Mbps)

Backbone Edge
Routers

Access-POPs

**Figure 4–14**    When using DPT, both Access-POPs and backbone edge routers are combined in a dual counter-rotating ring

There are several advantages of using DPT in the distribution layer. First, there are fewer fiber connections and interfaces required in both the Access-POPs and backbone edge routers because a ring topology is deployed. Second, DPT provides very good bandwidth efficiency through the spatial reuse capability and SRP-fa. Furthermore, high resilience is provided through IPS.

The major disadvantage is that a capacity upgrade is possible only for the whole ring and cannot be done partially.

## Optical Transport Infrastructure Implementation

How DWDM technology can be applied in the optical transport infrastructure and what kind of constraints have to be taken into account when designing DWDM-based networks are described in the following sections.

### DWDM SYSTEM BUILDING BLOCKS

The optical transport infrastructure is commonly deployed with unidirectional DWDM systems. These systems use all wavelengths carried over a fiber for transmission in one direction.

DWDM networks consist of terminal sites and optical line amplifier (OLA) sites. As already covered in Chapter 3, "Optical Networking Technology Fundamentals," ***terminal sites*** of DWDM systems consist of three basic building blocks: transponders, multiplexer/demultiplexer and booster amplifiers.

Figure 4–15 shows the building blocks of a DWDM p-t-p system for one direction only. For full-duplex operation, the left and the right terminal site have both a multiplexer and a demultiplexer; thus, both sites have a transmit and a receive multiwavelength trunk.

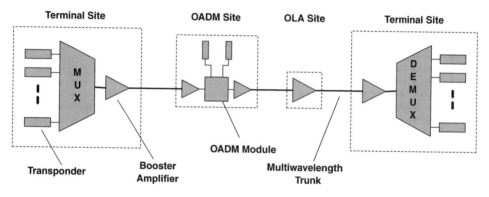

**Figure 4–15**    DWDM system building blocks

Transponders provide connectivity to gigabit routers and other service layer equipment. The multiplexer combines the wavelengths to a single optical signal, which is then amplified by a booster amplifier. At the receiving side, another booster amplifier again amplifies the optical signal, which is then separated by the demultiplexer into several optical channels.

As opposed to p-t-p systems, terminal sites of DWDM ring systems have two transmitting and two receiving multiwavelength trunks to provide full-duplex interconnectivity to the east and west neighbors in the ring. The multiplexers and demultiplexers are merged together into an optical ADM (OADM). The client signals from the transmit transponders are inserted by the OADM into the optical signal to be transmitted to the east neighbor terminal site or into the signal to be transmitted to the west neighbor terminal site. From the two received optical signals (east and west side), some wavelengths are also extracted and forwarded to the receive transponders.

Some DWDM vendors separate WDM operation for the wavelength range from 1529 to 1602 nm in two or three bands in their DWDM systems. The bands are separately multiplexed and demultiplexed, allowing greater modularity and scalability when upgrading the number of channels. The bands are also separately amplified, making it possible to optimize each optical amplifier for narrower and flatter regions, thus greatly reducing equalizing and tilt effects and reaching the best use of the optical amplifier's spectrum characteristics.

*OLA sites* are used along the fiber connection between the terminal sites to achieve very long distances. The amplifiers within the OLA sites compensate the attenuation losses introduced in the fiber. Some OLA sites can also be enhanced to provide add/drop functionality. An OADM module is inserted between the input and output booster amplifier. This OADM module extracts and inserts some specific wavelengths and terminates them at transponders. Some OADM modules allow dynamic wavelength selection for add/drop, whereas some support only static wavelength add/drop.

### TRANSPONDERS

Transponders are used to convert the signal from the service layer device to the exact WDM channel wavelength. They support either standard short-haul 1300-nm or standard long-haul 1550-nm optical interfaces, or both of them.

Most DWDM systems use transponders operating at one or more specific wavelengths. By inserting, for example, a single-channel transponder into the DWDM terminal site, one specific, fixed wavelength is available. There are two different types of transponders available: service-specific transponders and service-independent transponders.

*Service-specific transponders* require a certain type of framing and a certain line rate. These transponders perform 3R functionality. This means that they convert the optical signal into the electrical domain, perform signal reshaping and clock recovery, then convert the signal back into the optical domain. Most common service-dependent transponders are SONET/SDH transponders working at line rates of OC-12/STM-4, OC-48/STM-16, and OC-192/STM-64. By requiring SONET/SDH framing, bit error monitoring can be performed to monitor the signal condition and initiate optical protection functions in case of signal degrade.

*Service-independent transponders (transparent transponders)* provide only 1R functionality; thus, only the wavelength is transformed from the standard 1310-nm or 1550-nm client wavelength to the specific WDM channel wavelength. Transparent transponders are not as complex and are, therefore, a lot cheaper than service-specific transponders. Furthermore, any client signal, such as POS, DPT, Gigabit-Ethernet, or ATM, can be transmitted because these transponders are completely bit rate- and framing-independent. The disadvantage is that optical power budget calculation becomes more complex because the transponders do not provide a clear demarcation point. Furthermore, optical protection switching can be performed only according to optical signal power.

Most WDM systems also support the option of not using transponders. Instead, service layer devices may be directly connected to the multiplexer of the WDM terminals with *colored interfaces*. These interfaces are already providing the specific WDM channel wavelength. Of course, this wavelength must be very accurate and stable to avoid channel interference.

### WDM NETWORK DESIGN

When designing the optical transport infrastructure, several points, such as topology, channel spacing, maximum line rate, optical power budget, and upgrade ability, have to be considered.

#### TOPOLOGY

In general, ring and p-t-p topologies can be deployed at the optical layer. Point-to-point systems are mainly used within backbones to build high-bandwidth interconnects. The topology of the IP service layer might be star, ring, or mesh style, using the DWDM p-t-p links to interconnect adjacent locations. The main drawback of using p-t-p systems is that pass-through traffic is processed at the IP service layer.

By deploying a DWDM ring, a small number of fiber connections is needed to interconnect all locations. The network designer has the opportunity to provide any topology to the IP service layer. In particular, in cases where traffic patterns are fully meshed, tremendous simplicity can be achieved.

Figure 4–16 shows the ways that a network with five locations and a fully meshed IP service layer can be deployed.

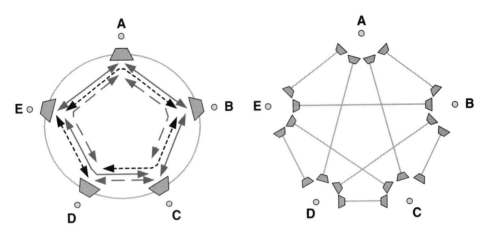

**Figure 4–16**    Flexibility and simplicity of DWDM ring networks

When using p-t-p systems, an extra DWDM terminal has to be added at both ends of the desired IP service layer connection. Within the ring implementation, only two additional transponders have to be inserted into the terminal site of the locations to be connected. A comparison of the required equipment for implementing the DWDM network is shown in Table 4–1.

**Table 4–1**    DWDM Equipment Requirements for Implementing a Fully Meshed IP Service Layer

|  | POINT-TO-POINT SYSTEMS | RING SYSTEMS |
|---|---|---|
| No. of DWDM terminals | 20 | 5 |
| No. of wavelengths per link | 1 | 3 |
| No. of fiber connections | 10 | 5 |

*CHANNEL SPACING*

The International Telecommunication Union ITU-T standardized a 50-, 100-, and 200-GHz channel spacing to ensure vendor interoperability. The 200-GHz channel spacing is commonly used for metropolitan WDM solutions. The 50- and 100-GHz spacing is used for long-haul DWDM systems.

The number of channels required in the initial deployment phase is, in most cases, quite low. Thus, DWDM systems could be configured with multiplexers and demultiplexers operating at 100-GHz channel spacing in the beginning.

If there is a future demand for channels requiring 50-GHz spacing, the DWDM systems cannot be upgraded "in service." Systems must be shut down, 50-GHz multiplexers and demultiplexers must be inserted, and additional transponders must be connected. Thus, it must be clear during the network design whether future channel requirements require 50-GHz spacing. If so, 50-GHz components must be used for the initial deployment.

### MAXIMUM LINE RATE

If very high line rates (e.g., OC-192/STM-64 at 10 Gbps) are desired, dispersion compensation has to be applied intensively. First, dispersion-compensating fibers, such as nonzero dispersion fibres (NZDs) and dispersion-shifted (DS) fibres, are to be used. In addition, special dispersion compensation units must be included in the DWDM systems to ensure proper optical transmission.

Another way to overcome dispersion limitations is to use 3R regeneration. Regeneration modules are inserted into intermediate OLA sites. Thus, the optical DWDM signal is converted into the electrical domain, reshaped and retimed, and converted back to the optical domain.

### OPTICAL POWER BUDGET

A key point in the design of DWDM networks is to calculate optical power budgets based on estimated optical power levels throughout the whole network. Each optical transmitter is designed to operate with an approximately constant output power. An output power range is specified where transmit power can vary.

Each optical receiver is designed to operate with an acceptable bit error rate (typically 10E-12) if the input power is within the range between the maximum receiver overload power and the minimum receiver sensitivity power. The difference between the minimum transmit power and the minimum receiver sensitivity is known as the *optical power budget*.

To ensure a reliable network design, the power budget must be greater than or equal to the sum of losses on the way from the optical transmitter to the receiver plus a design margin. These are all connector/splice losses, fiber attenuation, and power penalties. To avoid optical receiver damage, a minimum span attenuation must also be ensured. Figure 4–17 illustrates this calculation.

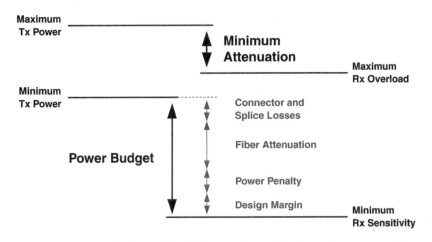

**Figure 4–17**   Power budget and minimum attenuation for a fiber connection

To compensate for an increasing error rate due to noise and interference (such as dispersion), some additional power must be added to the transmit signal. This amount of power is denoted as ***power penalty***. Commonly, the power penalty is included as part of the optical interface specifications.

A ***design margin*** is included to compensate for possible loss increases in the future. The transmit power, receiver sensitivity, and power penalties are typically specified as worst-case, end-of-life values by the equipment vendors. Thus, these components of the power budget calculation do not require a design margin. The design margin required, for example, for additional fiber splices to repair a broken cable. A typical value used is 1–3 dB.

Some typical values for the components used for calculating the power budget are shown in Table 4–2.

**Table 4–2**   Typical Values Used within the Power Budget Calculation

| COMPONENT | VALUE |
| --- | --- |
| Connector Losses | 0.2dB/Connector |
| Splice Losses | 0.01dB/Splice |
| Fiber Losses | 0.4 dB/km (at 1310 nm) <br> 0.3 dB/km (at 1550 nm) |
| Power Penalties | 1-2 dB |
| Design Margin | 1-3dB |

The design of the whole DWDM network is quite complex because many parameters have to be taken into account. Figure 4–18 shows how the optical path between two service layer nodes across the DWDM network is designed. To ensure reliable DWDM transmission, the optical interface specifications of routers or switches first have to be compared with the specifications of the DWDM transponders. Not just the operating wavelength must fit. In addition, the power budget between the service layer equipment interface and the WDM transponder interface must be correct.

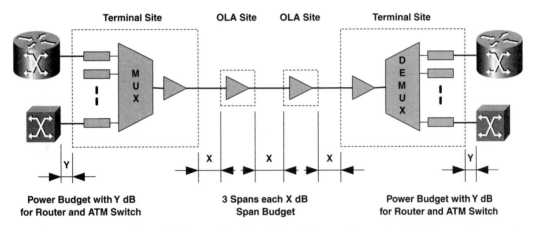

**Figure 4–18**   The optical power levels of the whole end-to-end path between two service layer nodes must be verified

The second part is the proper design of the path from the transponders of the transmitting terminal site to the transponders of the other receiving terminal site. DWDM vendors typically indicate the optical power budget between the sites via span budgets. Depending on the type of transponders used and the number of OLA sites being deployed, a specific *span budget* per fiber link (span) is defined.

## Restoration

To provide network reliability in static optical overlay networks, several protection and restoration functions can be applied at the service layer and/or at the optical transport layer. Resilient multiwavelength optical networks commonly use the fast protection functions of the optical layer. The DWDM equipment at the optical layer does not have any networking intelligence. Therefore, it is simply providing transport functionality. At the service layer, enhanced restoration functions are used to provide reliable end-to-end services.

## Optical Protection

Protection mechanisms at the optical transport layer are very limited in multi-wavelength optical networks, when compared with protection mechanisms provided by intelligent optical networks. Optical restoration functions are limited to optical p-t-p or ring subnetworks.

The major role of optical protection in multiwavelength networks is to provide reliable p-t-p connections used to interconnect the service layer nodes. End-to-end restoration is delivered through service layer restoration.

As ITU defined three optical layers in the "Architecture for Optical Transport Networks" in its recommendation G.872, optical protection can also be applied at three different layers.

At the transmission section layer, line protection can be used. At the multiplex section layer, commonly used protection mechanisms of SONET/SDH are applied. At the optical channel layer, optical channel protection is used (Figure 4–19).

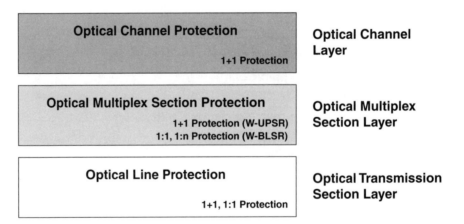

**Figure 4–19**    Protection in the OTN architecture

### OPTICAL LINE PROTECTION

Most p-t-p or ring WDM implementations at least provide optical layer survivability at the transmission section layer with simple 1 + 1 or 1:1 line protection, restoring all channels at a time.

When using 1 + 1 protection, the whole DWDM signal is protected via splitting up into two signals in the DWDM node and transmitting over two separate fibers (Figure 4–20). At the receiving node, both signals are compared, and the signal with the better optical signal-to-noise ratio (OSNR) or bit error rate (BER) is chosen.

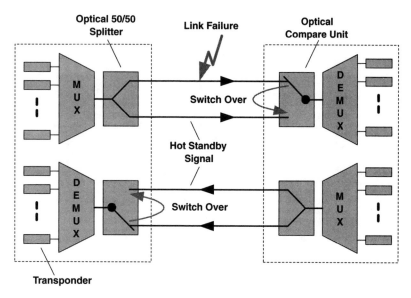

**Figure 4–20**     Optical 1 + 1 protection

When using 1:1 protection, the DWDM signal is sent over only one fiber at a time. If the working fiber fails, the DWDM signal is switched onto the protection fiber (Figure 4–21).

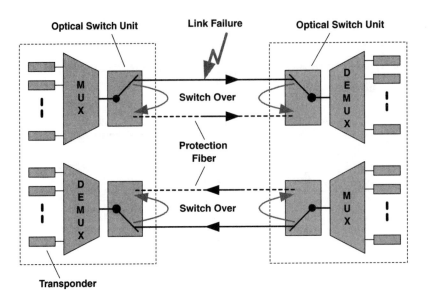

**Figure 4–21**    Optical 1:1 protection

With these approaches, all optical channels are protected against fiber cuts. The achievable protection time is lower than 10 ms.

### OPTICAL CHANNEL PROTECTION

The next step is to use an optical channel protection unit in conjunction with one working and one protection DWDM terminal to provide 1:1 protection for optical channels on a channel-by-channel basis at the optical channel layer.

This protection unit is composed of a transmission section and a receiver section. On the transmission section, the incoming optical client signal is split by a 50/50 coupler and sent to the transmit transponders of the working terminal and protection terminal (Figure 4–22). On the receiver section, the incoming signals from the receive transponders of the working line and protection line enter in a 1x2 optical switch unit (OSU) that makes the selection in case of failure within the optical layer. The selected channel is then sent to the client receiver.

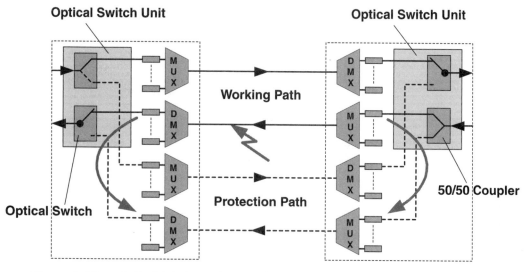

**Figure 4–22**   Optical 1:1 channel protection

The switching criteria might be based on a signal generated by the receive transponder, due to an input data loss. The criteria could also be the optical OSNR. The maximum recovery time from the failure occurrence to the complete recovery of the optical link is far less than 50 ms.

The major advantage of optical channel protection is that optical channels are not only protected against fiber cuts but also against multiplexer/demultiplexer (MUX/DEMUX) and transponder failures.

### OPTICAL MULTIPLEX SECTION PROTECTION

The most complex optical protection mechanism is optical multiplex section protection, where SONET/SDH restoration functions, which have been discussed in Chapter 3, "Optical Networking Technology Fundamentals," are implemented in DWDM ring systems. Reconfigurable DWDM ADMs provide dynamic wavelength channel allocation and protection switching to protect optical channels against network faults.

As described in the whitepaper "WDM Optical Network Architectures for a Data-Centric Environment" [TELL-1] and also outlined in the section "Network Survivability Principles" of Chapter 3 of this book, it can be distinguished between Optical Unidirectional Path Switched Rings (O-UPSR) and Optical Bidirectional Line Switched Rings (O-BLSR).

The **O-UPSR** architecture uses a two-fiber, counter-rotating ring configuration. One fiber is dedicated for working wavelengths, and one is used for protection wavelengths. An O-UPSR uses 1 + 1 protection. A wavelength on the working and protection fiber is allocated and transmitted for each channel. The receiving side compares both optical signals and takes the one with the better OSNR.

A logical block-diagram of a reconfigurable OADM utilizing 1 + 1 protection in an O-UPSR is shown in Figure 4–23.

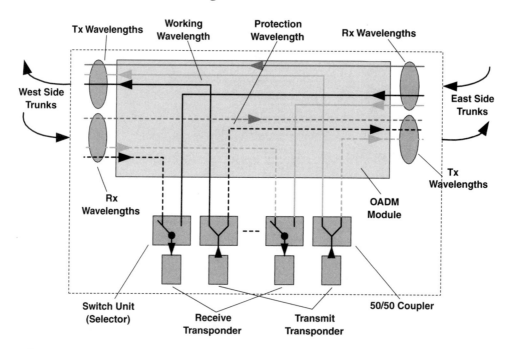

**Figure 4–23**    Reconfigurable OADM for 1 + 1 protection used in O-UPSRs

The transmit transponders are connected to 50/50 couplers, and each channel is split into a working and protection wavelength, which then are added to the DWDM signal in the ring by the OADM module. The receive transponders are connected to optical switch units, which select either the working or protection wavelength dropped by the OADM module.

The **O-BLSR** architecture uses either a two- or four-fiber counter-rotating ring configuration. Within the two-fiber O-BLSR, some wavelengths are used for allocating working channels; the remaining part is used as shared protection capacity. If a failure occurs, the reconfigurable ADM switches the failed wavelength onto a protection wavelength on the alternate path.

If the number of working wavelengths equals the number of protection wavelengths, the O-BLSR uses 1:1 protection. A more capacity-efficient approach is to use 1:n protection, where less than half of the wavelengths are shared across the rest of the wavelengths for protection (for example, 1:3 protection in a 64-channel DWDM system, where 48 wavelengths are protected by 16 wavelengths).

Supposing 1:1 protection, there are two possible wavelength allocation schemes. The first one is to use the same wavelengths for protection on both fibers. The second one is to use, for instance, the lower half of the wavelengths for protection on the fiber in the clockwise direction and the other half of the wavelengths on the fiber in the counter-clockwise direction.

The advantage of the latter one is that protection switching can occur without wavelength conversion in the OADM. Wavelength conversion may be done fully optical in the future but will commonly be done via electro/optical conversion. This advantage might not be worth too much in long-haul backbone applications, where electro/optical conversion is required in any case to regenerate the optical signals.

## Service Layer Restoration

Service layer equipment, typically routers, is capable of numerous protection and restoration functions. Within multiwavelength networks especially, IP restoration is used to implement resilient end-to-end network connectivity.

### AUTOMATIC PROTECTION SWITCHING

Automatic Protection Switching (APS) used in SONET networks or Multiplex Section Protection (MSP) used in SDH networks might also be used within optical networks. DWDM terminals could provide a working and a protection SONET/SDH-compliant interface (Figure 4–24). In case of a failure, an interface switchover can be triggered through APS/MSP. This requires that the transponders of the DWDM equipment are processing the SONET/SDH overhead bytes used by APS/MSP.

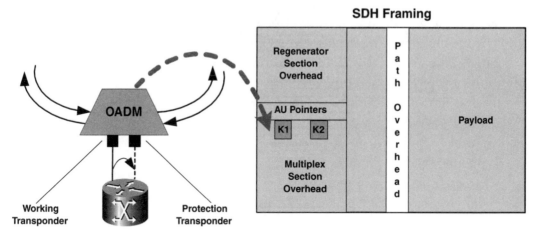

**Figure 4–24**    Interface protection with Automatic Protection Switching (SONET) or Multiplex Section Protection (SDH)

Such applications might be very common in network implementations, where new-generation SONET/SDH equipment with integrated DWDM functionality is used. APS/MSP protection has two major disadvantages. The first is that 50% of the capacity (one whole interface) is wasted for protection, and the second is that the interface switchover still requires routing protocol reconvergence.

### INTELLIGENT PROTECTION SWITCHING

DPT dual-fiber self-healing rings provide enhanced resilience because of the IPS functionality of SRP, the layer-2 MAC protocol used by DPT equipment, which is described in detail in Chapter 3, "Optical Networking Technology Fundamentals." In case of link or node failures, IPS initiates ring wraps and transmits traffic onto the protection fiber (Figure 4–25).

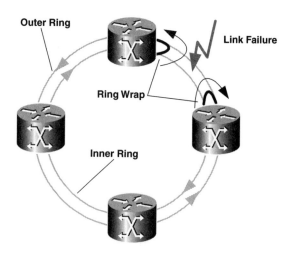

**Figure 4-25**   Wring wraps performed by IPS

IPS can be seen as the counterpart of APS/MSP used in SONET/SDH networks; however, there are several advantages over APS.

IPS does not activate or deactivate certain channels or fibers but, like DPT, is a packet-based technology. IPS efficiently redirects statistically multiplexed traffic. It integrates layer 2 and layer 3 into the protection mechanism, as opposed to APS, which operates completely independent from layer-3 routing. Through the integration of layer 3, IPS has access to layer-3 protocol information. With this, IPS is able to deliver guaranteed bandwidth for voice/video/mission-critical applications and to provide remaining bandwidth for best-effort traffic.

Because the achievable protection time with IPS is below 100 ms, no routing protocol reconvergence is required. Fiber cuts can be completely handled at layer 2. For a detailed description of the IPS protection mechanisms, review Chapter 3, "Optical Networking Technology Fundamentals."

### IP RESTORATION

#### *APPLICATION-BASED RESTORATION*

Layer-3 restoration delivers some distinct benefits. High flexibility is provided when provisioning restoration paths and when specifying certain restoration levels.

For example, voice traffic—which is very delay- and jitter-sensitive—can be differentiated and restored at the highest restoration. On the other hand, best-effort Internet traffic may be restored at the lowest level, because it can be easily buffered and rerouted in the network.

Because of this, service providers are able to define certain traffic categories and can provide several options to their customers.

### LAYER 3 ROUTE CONVERGENCE

Aside from the advantage of flexibility, there is one big disadvantage: the restoration time of standard routing protocols, such as Open Shortest Path First (OSPF), Intermediate System to Intermediate System protocol (IS-IS), or Border Gateway Protocol (BGP), is typically in the range of several seconds. Timer and route-dampening mechanisms provide network stability but lead to very long convergence times.

For example, when looking at OSPF, there is a Hello protocol used to discover and maintain neighbor relationships. Each router is sending periodically, every 5 or 10 seconds, a Hello message to its neighbors. After not receiving a Hello message for a couple of times, the router realizes that there has been a failure on the link to its neighbor.

The first approach to crank down the layer-3 restoration time is to decrease the routing protocol timer values. By doing this, the convergence time can be tuned down to 1 or 2 seconds. Of course, this leads to a higher amount of traffic produced by the routing protocol itself. As the link bandwidth has increased dramatically, this is no longer a problem. The influence of increased ability for route flapping and routing protocol instability can also be considered to be low because links and network equipment have become a lot more resilient.

Another possibility might be that the router interfaces use layer-1 or layer-2 mechanisms to detect link failures and immediately report a failure to the routing engine. An example for this might be APS/MSP used in SONET/SDH networks.

### LOAD BALANCING

Load balancing is a very common technique to add redundancy into a network. By using load balancing, parallel paths are used to split up traffic and transmit on both links simultaneously. To facilitate load balancing, the routing protocol is tuned by assigning certain metric values to the links throughout the network. As a result, there are parallel paths with equal metrics between the source and destination; therefore, the routing protocol has two routing table entries for the same destination.

Figure 4–26 shows how load balancing can be used to protect traffic flowing between routers A and D and between routers B and E. In case of a fiber cut, the routing protocol simply cancels the entry for the failed link and flushes its routing table. No SPF calculation is required, which leads to a very short restoration time. During this time period, only 50% of the traffic is affected.

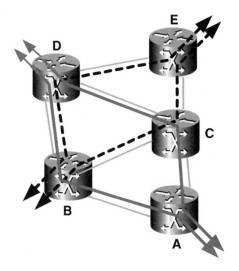

**Figure 4–26**     OSPF load balancing through a meshed core

### MPLS-TE FAST REROUTE

One very good way to implement restoration functionality at the service layer is to use the restoration mechanisms of MPLS-TE, which gives the flexibility to specify what paths, links, and/or nodes should be protected. Furthermore, what types of traffic to protect can be defined, as well as how much resources need to be reserved for the backup path.

MPLS-TE provides path protection and local link/node protection. Path protection shown in Figure 4–27 uses an end-to-end backup path to restore the traffic in a failure situation. The achievable restoration time can be equal to or greater than 1 second. For critical or real-time traffic, local protection of certain links or nodes can be used to route traffic around a network failure within less than 50 ms.

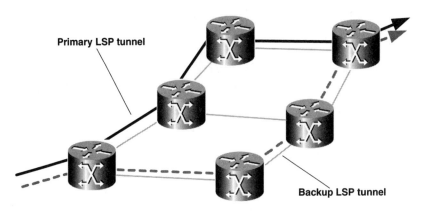

**Primary LSP tunnel**

**Backup LSP tunnel**

**Figure 4–27**  An active and backup path is set up by using MPLS label switch path tunnels (LSP tunnels)

MPLS-TE defines the term ***traffic trunk***, which represents a group of traffic flows having the same ingress and egress node in the MPLS network. Each traffic trunk has several trunk attributes assigned. One of them is the resilience attribute, which determines how the traffic should be protected. Using the resilience attribute, several restoration policies can be flexibly specified in order to handle network failures.

For details on MPLS-TE and the provided protection mechanisms, refer to Chapter 3, "Optical Networking Technology Fundamentals."

## Dynamic IP Optical Overlay Control Plane

Wavelength provisioning is the key issue for static multiwavelength networks, which were covered in the previous section. This issue is to be solved within wavelength routing networks. IP is the "traffic of choice"; thus, optical cross-connects (OXCs) are combined with IP routing intelligence to control wavelength allocation, setup, and tear-down dynamically.

The result is a much more scalable network with an expedited provisioning process and enhanced restoration capabilities. Any virtual topology can be provided. Dynamic, data-driven wavelength provisioning ensures efficient network utilization. To adapt to changing traffic patterns, the limited number of wavelengths processed concurrently are rearranged automatically or on manual request.

To avoid misinterpretation, the term *OXC* does not necessarily mean full optical switching. Most OXCs terminate the optical signals at the input port, switch the electrical signals through an electrical switch fabric, and convert the signals back to the appropriate wavelength.

## Wavelength Routing Overlay Model

Several vendors already provide proprietary wavelength routing solutions. This enables service providers to build an intelligent OTN, as opposed to a static OTN, which simply consists of DWDM p-t-p and ring subnetworks. A wavelength routing network consists of a mesh of p-t-p DWDM links and wavelength routers (WRs) at the junction points.

All currently available wavelength routing solutions concentrate on the optical transport layer. Well-known IP routing protocols, such as OSPF and IS-IS, or ATM routing protocols, such as private network-to-network interface protocol (PNNI), are adapted to create a routing protocol to be used by the WRs. Using this wavelength routing protocol, connections can be dynamically provisioned to interconnect IP routers or other service layer equipment that resides on top of the OTN. Because the wavelength routing protocol is the only protocol running on the WRs, the IP network does not participate in the wavelength routing process and interacts with the OTN in a client/server relationship. This network model (where two independent network layers are deployed) is called an *overlay model* and is shown in Figure 4–28.

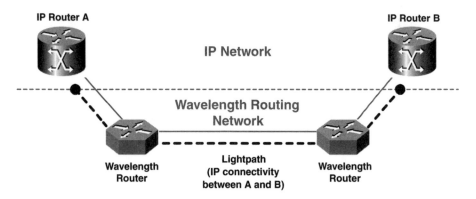

**Figure 4–28**     In the overlay model, the wavelength routing process of the OTN is completely independent from the service layer

Currently, all WRs must be from the same vendor to be able to participate in the routing process. To ensure vendor interoperability that allows a mix of different WRs to be used in the OTN, an architectural model has to be standardized. Several standardization bodies have already started development in that area.

A framework for IP transport over optical networks has already been proposed within the Internet Engineering Task Force (IETF) draft "IP over Optical Networks—A Framework" [IETF-11]. Details on requirements and mechanisms for bandwidth management and restoration in wavelength routing networks are described in the IETF draft "Control of Lightpaths in an Optical Network" [IETF-12]. The considerations about the architectural model described in this section are based on these two drafts.

## Architecture and Elements

The key element of wavelength routing networks is the *wavelength router* (WR). It is an IP router combined with an OXC used as interconnection node in meshed optical networks. Service layer equipment such IP routers and ATM switches interfacing the optical network are referred as *wavelength terminals* (WTs).

A wavelength router is attached via multiple *network ports* to DWDM terminals, which provide DWDM connectivity to other WRs. The WR's *drop ports* are used to connect the WR to service layer equipment. Network and drop ports are single wavelength interfaces commonly operating at 1,300 nm or 1,500 nm. In case of DWDM-enabled service layer equipment, DWDM systems are connected to drop ports to provide a DWDM signal to the attached router, ATM switch, or SONET/SDH terminal.

There are also wavelength routers with integrated DWDM terminals. In this situation, the above-mentioned network and drop ports are only virtually existent and represent the wavelengths of the DWDM signals.

The typical architecture is shown in Figure 4–29.

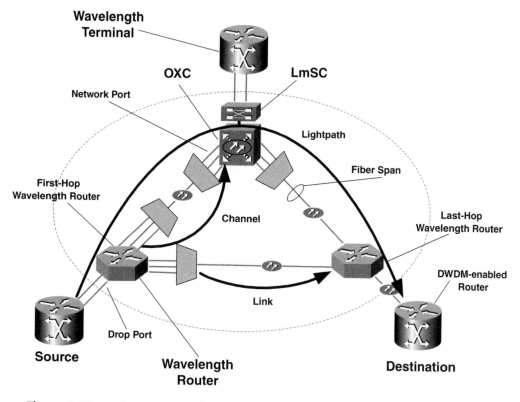

**Figure 4–29** The elements of an intelligent optical network

According to the standardized OTN defined in the ITU-T recommendation G.872, naming of some network elements should be outlined.

The physical interconnection between neighboring DWDM terminals consisting of a couple of fiber cables is defined in G.872 as Optical Transmission Section trail (OTS trail). As is also done in this work, an OTS trail is often referred to as a *fiber span*. DWDM systems utilize multiple wavelengths over a fiber span. A *link* is defined as a set of channels between two network nodes where each *channel* is a unidirectional optical tributary connection. In general, a channel represents a certain wavelength. If the WR facilitates electrical multiplexing, multiple channels are transmitted over a single wavelength. For example, four STM-16 channels are multiplexed onto a wavelength at STM-64 line rate. The corresponding ITU-T naming for all channels carried over a fiber span is defined as an Optical Multiplex Section trail (OMS trail) within G.872.

Some similarities can be found when comparing wavelength routing networks providing TDM functionality with ATM networks, as shown in Figure 4–30. In ATM networks, logical connections are defined by using virtual path identifiers (VPIs) and virtual channel identifiers (VCIs). In wavelength routing networks with TDM functionality, logical connections are defined through the use of wavelengths (wavelength = VPI) and channels (TDM timeslot = VCI).

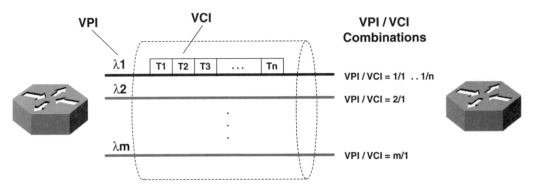

**Figure 4–30**    Comparing ATM to wavelength routing networks with TDM functionality

The wavelength router dynamically switches channels between input and output ports to deliver an end-to-end optical connection through the OTN terminated at the edge. End-to-end connectivity is defined in G.872 as an optical channel trail (OCH trail). There are two different kinds of OCH trails. In an optical network without wavelength conversion, the same wavelength has to be allocated to the OCH trail throughout the whole network. In this case, the OCH trail is called a *wavelength path* (WLP). If wavelength conversion is provided, different wavelengths can be used for the OCH trail, and we talk about a *lightpath*.

A lightpath is a unidirectional fixed bandwidth connection set up via the allocation of a channel on each span along the desired path through the optical network. A bidirectional lightpath is established by using two lightpaths in opposite directions. Although IP routing is used to coordinate lightpath provisioning, lightpaths can also carry non-IP traffic, such as ATM or SDH.

Similar to the ATM architecture, two different interfaces are defined. WRs are interconnected via network ports/trunks, providing an *Optical Network-to-Network Interface* (O-NNI). Wavelength terminals are attached to wavelength routers via the drop ports/trunks, providing an ***Optical User-to-Network Interface*** (O-UNI).

## Wavelength Routing Control Plane

The control plane is responsible for establishing an end-to-end connection, also known as a ***lightpath***. A lightpath is then used to set up IP or other service layer protocol reachability. There are two ways of implementing an IP-based control plane used in a wavelength routing network.

The first one is to attach external IP routers via a standard control interface to each OXC. These routers may be referred as ***wavelength routing controllers*** (WRCs) and provide functions such as optical resources management, configuration and capacity management, addressing, routing, traffic engineering, topology discovery, and, of course, restoration.

The control interface specifies a set of primitives used by the WRC to configure the OXC. The interface primitives include:

- ***Connect:*** To cross-connect a channel from an incoming link with a channel at the desired outgoing link
- ***Disconnect:*** To remove a configured cross-connection
- ***Switch:*** To change the incoming channel/link combination of a configured connection

Conversely, the OXC communicates with the WRC. A primitive example is:

- Alarm: Informing the WRC of a failure situation

An OXC vendor-specific device translates between the standard control interface primitives and the proprietary controls of the OXC.

The second way is to integrate IP routing functionality into the OXC, thus developing a single-box wavelength router. Both approaches are shown in Figure 4–31.

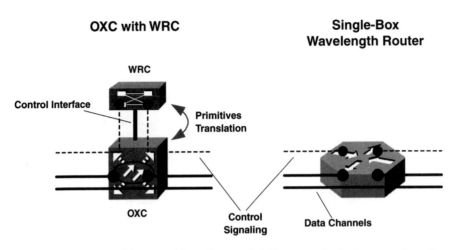

**Figure 4–31** WRC with control interface and OXC versus single-box wavelength router

The wavelength router maintains a cross-connect table used to distinguish how a data stream is switched from an input port to an output port to provide an end-to-end optical connection through the OTN. WRs with integrated DWDM terminals maintain a cross-connect table with entries having input port and input channel associated with an output port and output channel. This difference is shown in Figure 4–32.

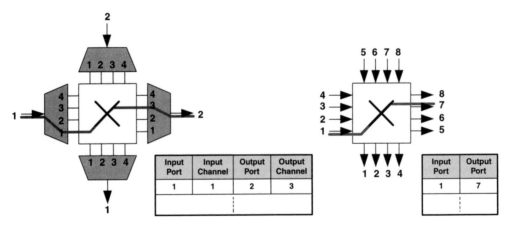

| Input Port | Input Channel | Output Port | Output Channel |
|---|---|---|---|
| 1 | 1 | 2 | 3 |
| | | | |

| Input Port | Output Port |
|---|---|
| 1 | 7 |
| | |

**Figure 4–32** Cross-connect table of wavelength routers with or without integrated DWDM terminals

The control plane exchanges control traffic through a Digital Communication Network (DCN). This DCN may be an in-band or out-of-band control network. For in-band DCN connections between wavelength routers, a default-routed (one-hop) lightpath is set up on each link to other WRs. As can be seen in Figure 4–33, the first channel on each link is allocated for in-band management. Control messages encapsulated in IP packets are sent over this default-routed lightpath.

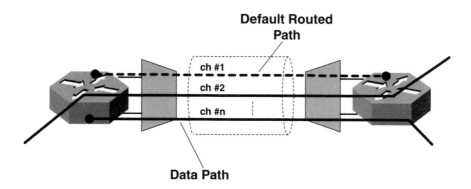

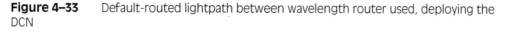

**Figure 4–33** Default-routed lightpath between wavelength router used, deploying the DCN

For in-band DCN connections to wavelength terminals with DWDM interfaces, a default-routed lightpath is also used. If WRs are connected with single wavelength interfaces to the service layer, the in-band DCN connection is implemented through IP packets transmitted within framing overhead of the service layer equipment interface.

A second approach is to deploy an out-of-band DCN. Routers and leased lines are used to set up a completely separated IP network interconnecting all wavelength routers and wavelength terminals.

Across the DCN, the IP routing protocol instance of each wavelength router exchanges information about the topology and status of the OTN. An enhanced IP routing protocol is commonly used for wavelength routing; every network element must have a unique IP address assigned to be addressable. This includes the wavelength routers and their interfaces and any other network element in the OTN (such as optical line amplifiers of long-haul fiber spans).

Figure 4–34 shows an example where, to achieve clarity, the DWDM terminals between the wavelength routers are not shown. All WRs in the network have 10.1.0.x IP host addresses. Other equipment, such as the one optical line amplifier, has 10.2.0.x IP host addresses assigned.

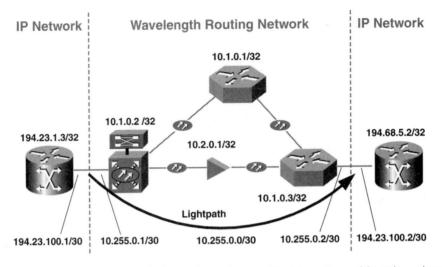

**Figure 4–34**    Every network element requires an IP address to enable IP-based control

The overlay model is typically used in the case where the OTN is owned by an optical interexchange carrier. Other service providers are buying lightpaths to establish their IP network at the service layer and are using their own IP address space for their service layer equipment. Thus, the left IP router has the IP host address 194.23.1.3, and the right IP router has the IP host address 194.68.5.2 assigned.

To make the lightpath between the two service layer boxes addressable in the wavelength routing domain, the WRs can have an IP address out of the IP sub-net 10.255.0.0/30 assigned to their corresponding interface.

The lightpath is providing a service layer IP connection between the two routers. This means that the corresponding IP router interfaces might also have IP addresses assigned.

## Lightpath Provisioning

The overlay model defines a client/server model where the service layer is the client of the optical transport layer. Two independent control planes, one for the service layer and one for the optical transport layer, are deployed. Therefore, routing and signaling are completely separated. The optical transport layer provides p-t-p lightpaths to the service layer. The lightpaths provide next-hop connectivity between the service layer nodes and deploy the desired virtual topology used for

IP transport. The lightpaths through the OTN may be set up statically through manual provisioning or dynamically by the wavelength routing mechanism.

The wavelength routing protocol might be a proprietary solution, whereas a multivendor situation requires a standard protocol. A standard wavelength routing protocol will be Multiprotocol Lambda Switching (MPLmS), which is already under development by the IETF. MPLS is used for lightpath routing and service provisioning in an OTN. In an overlay model, one MPLmS control plane would be used to control lightpath provisioning in the OTN, and a completely separated MPLS control plane would make use of this connection to provision label-switched paths (LSPs) in the router network of the service layer. MPLmS is covered in detail in a later section of this chapter, called "Integrated IP Optical Peer Control PlaneIP Integrated with the Optical Layer in the Peer Model."

When remembering the issues already experienced with integrating IP and ATM networks, it becomes clear that the overlay model also has some drawbacks when used in the optical network field. IP routing requires mesh peering between the routers at the service layer. So the $N^2$ problem that appeared in IP over ATM networks also has a significant impact on optical networks designed after the overlay. Hence, the overlay model is not the perfect solution in situations where the optical transport layer and the service layer are owned by a single service provider.

A typical situation where the overlay model is considered is when service providers lease their transport infrastructure from optical exchange carriers or bandwidth providers. In this case, we have two completely separate administrative domains—one for the OTN and one for the service layer—each requiring its own control plane.

An advantage of a proprietary wavelength routing solution and the overlay model is that some vendors of WRs implement some enhanced functionality into their proprietary wavelength routing solutions. This can include, for example, very fast protection mechanisms or advanced QoS functionality. Using the proprietary solution for wavelength provisioning in the OTN and combining it with an overlaid MPLS control plane for the service layer would create a good solution with an overlay model.

## Lightpath Attributes

WRs are requesting lightpaths to be established or torn down as connectivity at the service layer, commonly for IP traffic, is required. Several attributes must be defined in these requests.

First, each lightpath is represented via a unique *lightpath identifier*. Its capacity is defined by the *bandwidth* attribute, which is restricted to two possible values. Optical networks are to provide high-bandwidth connections for the upper layers. Current switching matrixes used in wavelength routers are optimized for switching OC-48/STM-16 or OC-192/STM-64 channels; thus, possible bandwidth values are 2.5 or 10 Gbps. As optical technology evolves, higher bandwidth values will also be possible.

As is the nature of an optical connection, a lightpath is unidirectional. For a full-duplex data flow, a bidirectional connection is required; thus, two unidirectional lightpaths in opposite directions have to be requested.

By classifying lightpaths in several *restoration classes*, reactions to network failures can be defined on a per-lightpath basis.

## Centralized Lightpath Routing

As shown in Figure 4–35, route calculation for lightpaths can be centralized by using a traffic engineering server. WRs act as the corresponding clients. The server maintains an information database, including topology and inventory of physical resources. Information about current resource allocations and the applied addressing and naming scheme is also maintained.

Algorithms on the server facilitate state maintenance and bandwidth allocation mechanisms. Algorithms for resource allocation management and reoptimization are also provided to adapt to changing network utilization. To ensure high network resilience, fault detection and recovery algorithms are implemented.

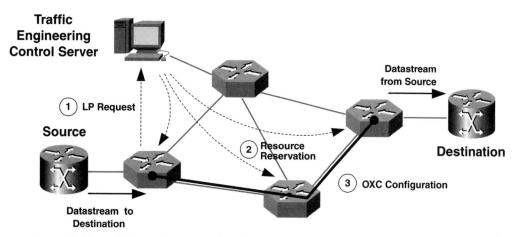

**Figure 4–35** When using centralized routing for lightpath establishment, the wavelength routers act as clients of the traffic engineering server

WRs request a lightpath to be set up at the server. The server checks resource availability and initiates resource allocations at each hop of the lightpath through the network. The centralized approach keeps network control and service provisioning very clear. In the case of network failure, centralized network restoration might not be possible within the required time constraints of optical networks.

## Distributed Lightpath Routing

Distributed network control, shown in Figure 4–36, ensures lightpath provisioning in very small timelines. This is especially important for network restoration. Each wavelength router maintains its own information database and its own set of algorithms.

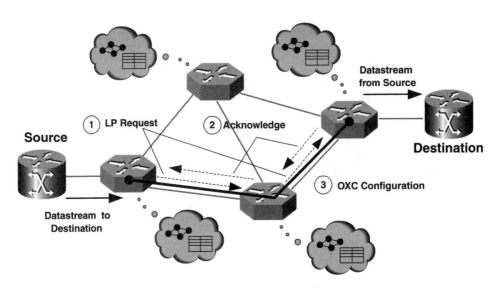

**Figure 4–36**    Lightpath establishment with distributed routing

WRs perform neighbor discovery after systems boot-up. Using the collected information, the wavelength router builds a topology map. The WR then retrieves information about the local resources of the OXC and creates resource hierarchies. Resource information is exchanged with all other WRs by extending the functionality of the IP routing protocol that the WR is running. Routing updates are flooded across the network. Each WR maintains a routing table and an OXC link resources information base. Constraint-based routing is used to define an appropriate path through the wavelength routing network.

Figure 4–37 shows a simplified high-level flow chart of the lightpath provisioning process in order to describe the principle procedures. Real-world implementations might differ slightly, according to the signaling protocol used. If a lightpath is requested to a certain destination, the first-hop wavelength router determines an explicit route through the network and initiates establishment by sending a setup message. The setup message travels across the desired path through the optical network. At each hop, the WR verifies whether enough resources are available. If yes, resources are allocated, and the setup message is sent to the next hop. If not, a setup failure message is sent back toward the first-hop WR.

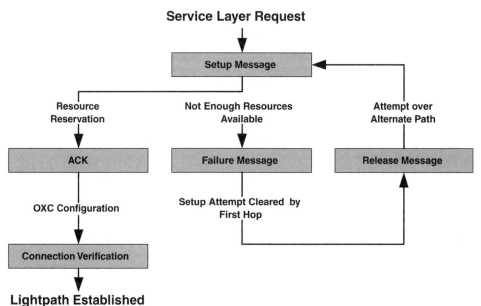

**Figure 4–37**    Simplified lightpath setup flowchart

After the setup message has been received by the last-hop wavelength router and resources have been allocated successfully at each hop, an acknowledgement is sent back toward the first-hop WR. The setup process is finished, and WRs configure their OXCs. Successful configuration and lightpath activation may be verified by tandem connection identification verification, as defined in the OTN architecture of G.872.

If a lightpath has to be removed, a release message is sent across the path, requesting each wavelength router to delete the resource allocations.

## Optical UNI Signaling

The Optical Internetworking Forum (OIF) has proposed a standard O-UNI, as described in Chapter 2, "Optical Networking Standardization."

With this standard O-UNI, *dynamic on-demand requests* are possible between the separated control planes of the optical transport layer and the service layer. The lightpath establishment is the same as described before, but the lightpath setup request is a dynamic one sent by the wavelength terminal. It is not manual via the network management of the WRs.

## Wavelength Conversion

Wavelength assignment is very simple when using electrical wavelength routers. These WRs have an electrical switching matrix and, therefore, provide full wavelength conversion. Furthermore, optical networks using electrical WR (also called *opaque optical networks*), provide 3R regeneration, including reshaping, retiming, and reamplification.

Within all optical networks (also referred to as *transparent optical networks*), optical wavelength routers are used, and wavelength allocation becomes more sophisticated. Most optical WRs are based on Microelectromechanical Systems (MEMS), which allow wavelengths to be switched between DWDM interfaces, but do not provide the abilities to connect between wavelengths. If this is the case, the same wavelength has to be allocated to the lightpath throughout the whole optical network. Because the number of available wavelengths per WR is limited, it may not be possible to establish a requested lightpath when the request is blocked. The blocking probability is often used as a metric to analyze the effectiveness of wavelength conversion.

The problem with establishing a lightpath in a transparent optical network is known as the *routing and wavelength assignment problem*. Details on this are out of the scope of this work but can be found in the book *Optical Communication Networks* [MUKH-1].

## Restoration

Data networks using standard IP routing protocols cannot provide the required level of resilience for current and upcoming applications. Voice integration and multimedia applications create the demand for the same level of "resilience," as it is known from telecommunication networks.

There are several restoration approaches for data networks. One is a centralized control server responsible for service provisioning and traffic protection. The achievable restoration time exceeds several minutes. The counterpart is a distributed architecture, where the restoration time ranges at about 2–10 seconds, depending on the routing protocol convergence.

SONET/SDH rings are used in telecommunication networks. The implemented protection switching functions restore network failure in less than 50 ms, avoiding any service interruption for upper-layer applications. The drawback of SONET/SDH rings is the tremendous bandwidth inefficiency because 50% of the capacity is reserved for protection purposes.

Mesh networks provide the most efficient way to protect working traffic in terms of bandwidth utilization because each link can provide protection for several different network failures. The achievable restoration time depends on the implementation of the distributed control protocol.

A wavelength routing architecture (as described before) can deliver a resilient, manageable, optical network using mesh restoration mechanisms. One of the key advantages to using wavelength routing in the OTN is that dynamically provisioned end-to-end paths through the network are automatically restored through rerouting the working traffic onto backup paths.

## Protection Strategy

Wavelength routing gives the flexibility to configure lightpaths with certain resilience and QoS attributes to meet the applications requirements.

The chosen *resilience attribute* assigns an appropriate protection mode to the lightpath. The configurable QoS level makes it possible to differentiate between several lightpaths and to prefer one to another.

### PRENEGOTIATED PROTECTION

Using this type of protection, both the primary and backup paths are provisioned at the same time. It is important that the backup path is completely node- and link-disjointed to prevent the backup path from being affected by the network failure; this would force the primary path to fail. Two different types of prenegotiated protection must be distinguished.

The first type of prenegotiated protection is 1 + 1 protection (shown in Figure 4–38). Both the primary and backup paths are set up at the same time, and the required bandwidth is allocated for both. The working traffic is sent over both paths. By doing this, the destination node can select independently from the source node which path to use for receiving traffic by monitoring the signal quality or bit-error rate. As a consequence, no signaling or information exchange is required.

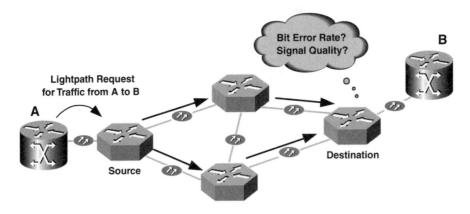

**Figure 4–38**   With 1 + 1 protection, working traffic is sent over two parallel and disjointed paths, and the destination node selects one of them

The second, more resource-efficient type is 1:1 protection (shown in Figure 4–39). When provisioning the primary path, the backup path is negotiated but the allocated bandwidth can be used for low-priority traffic. If a failure is detected by the adjacent nodes, they then propagate failure notifications up- and downstream along the path. After several ms, the source node detects the failure. It then sends signaling upstream to inform the destination node to use the backup path and drops the low-priority traffic to free up the allocated bandwidth for the backup path. The information exchange between the nodes required for signaling and failure propagation may use a certain signaling channel implemented in the layer-1 or layer-2 overhead or a broadcast mechanism to ensure fast enough information propagation throughout the network.

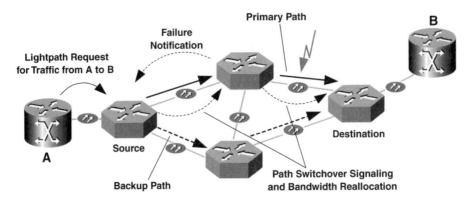

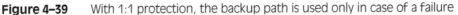

**Figure 4–39**   With 1:1 protection, the backup path is used only in case of a failure

A third type of protection is 1:n protection, which is a special case of 1:1 protection. The mechanisms behind them are the same but up to n lightpaths are sharing one backup path for protection.

Prenegotiated protection is used for mission-critical and high-priority traffic and delivers the best restoration time, which is below the 50-ms limit defined by SONET/SDH.

### ON-DEMAND PROTECTION

If this type of protection is used, there is no backup path established when provisioning the primary lightpath. If a failure occurs, the adjacent nodes detect and propagate the failure to the originating and terminating node. The originating node then initiates the setup for an alternate disjoint path and reroutes the traffic after the setup is completed.

Achievable restoration times are, of course, longer than with prenegotiated protection. They vary, depending on the implementation but can be expected in the range from 200 ms to 1 second. The major advantage of this type of protection is that the protection capacity is shared among the working lightpaths.

### QoS LEVELS

Through assigning QoS levels to lightpaths, the order in which lightpaths are to be restored can be assigned. Furthermore, one can also define whether one restoration process may preempt another lightpath.

Combined with the choice of several protection types, a network operator has the tools to develop a protection strategy required for delivering a set of different service classes for its customers.

## Advanced Protection Concepts

### LOCAL PROTECTION

Both local and end-to-end path protection are used to optimize network resilience. When using *local protection*, the node adjacent to the failure reroutes the lightpath around the failure. This can occur a lot faster because no failure notification across the network is necessary.

### HIERARCHICAL PROTECTION

Some wavelength routing protocols might be of a hierarchical nature. That means that the wavelength routing network can be segmented into several areas, as shown in Figure 4–40, in order to optimize the resilience and scalability of the network.

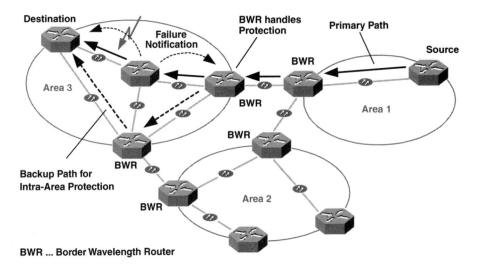

BWR ... Border Wavelength Router

**Figure 4–40**    Protection can be scaled through separating the wavelength routing network into multiple areas

In doing this, one must distinguish between inter- and intra-area protection because the network is separated into several areas. The wavelength routers interconnecting two or more areas are called **border wavelength routers**.

**Intra-area protection** implies that a failure can be restored within the affected area. This can either be done by local protection (i.e., by the node adjacent to the failure) or by the border wavelength router on the path to the source.

Normal end-to-end path protection across the whole wavelength routing network can be seen as **interarea protection**.

## Integrated IP Optical Peer Control Plane

The use of MPLS-TE control mechanisms for implementing an IP-based control plane for wavelength routing networks is under study and has been proposed within the IETF draft "Multi-Protocol Lambda Switching: Combining MPLS-TE Control with OXCs" [IETF-13]. How the MPLS-TE control plane can be adopted to be able to control OXCs is described in the IETF draft "Multi-Protocol Lambda Switching: Issues in Combining MPLS-TE Control with OXCs" [IETF-14]. Both drafts provide the background for the following description of MPLmS.

## Applicability Considerations

There are three major requirements to be fulfilled by the OXC control plane:

1. It must be able to establish lightpaths.
2. It has to support traffic engineering functionality to utilize available resources efficiently.
3. Restoration and protection mechanisms must be provided to ensure network resilience.

### Advantages

Using MPLS-TE for implementing a control plane delivers the ability to provision lightpaths in real-time across the OTN and service layer network. Because MPLS-TE has already been deployed and tested in the ATM world, no development of new control protocols for coordination between service layer network elements and optical transport layer network elements is required.

### Comparing WRs and LSRs

The first reason why MPLS-TE is an obvious solution for the OXC control plane problem is that WRs and LSRs are very similar in their architecture and functionality. Both LSRs and WRs have separated data and control planes; thus, switching of optical connection or forwarding of data traffic is completely separated from routing and signaling mechanisms.

A LSR provides unidirectional virtual p-t-p connections, called LSPs. Traffic belonging to defined Forward Equivalent Classes (FECs) is aggregated and transported over these LSPs. A WR provides unidirectional optical p-t-p connections called *lightpaths*, used to transmit traffic aggregated by connected service layer equipment.

The LSR maintains a table with Next-Hop Label Forwarding Entries (NHLFEs) and, therefore, knows what label to use to send traffic to each next hop. A WR maintains a cross-connect table. According to this table, the switching matrix is programmed to switch a channel through the WR to the desired neighboring WR.

In addition to the above listed similarities, there are two major differences. First, the LSR must process the packets for doing the label lookup. The WR does not perform any packet-level processing. It is switching channels, regardless of the transported traffic type and payload. Second, the switching information for WRs is the lightpath ID and not the label carried as part of the data packet.

### Comparing LSPs and Lightpaths

The second reason for using MPLS-TE is that lightpaths are very similar to LSPs. Both are unidirectional, p-t-p virtual paths between an ingress and an egress node. Although LSRs process the packet header, the payload is transparent to intermediate network elements for both LSR and WR networks. LSPs define a virtual topology over the data network, as do lightpaths over the OTN. Allocating a label to an LSP equals allocating a channel to a lightpath.

Figure 4–41 shows the two different meanings of the term *label* in the service layer MPLS architecture and the MPLmS architecture of the OTN.

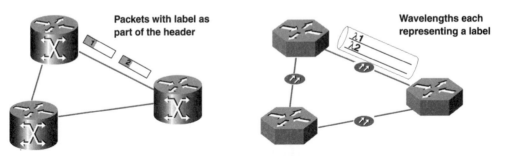

**Figure 4–41**　Definition of the term *label* in the MPLS and MPLmS architectures

In the MPLS architecture, the term *label* denotes a fixed-length value carried in the cell/packet header. LSRs process the packet header to distinguish between the different LSPs carried over the link.

In the MPLmS architecture, the term *label* denotes a certain wavelength on a fiber span or, if the WRs are TDM capable, a certain TDM channel of a wavelength. Thus, WRs process incoming traffic at the channel/port level and distinguish between different LSPs, according to their incoming port/channel identifier.

The available label space is not an issue when implementing an MPLS network with either IP routers or ATM switches. Looking at the OTN, the number of available labels per WR is defined by the number of wavelengths provided by the DWDM systems; the label space becomes a certain constraint. Normally, an MPLS network requires a label per FEC. In a typical provider network, there might be more than 60,000 routes in the routing table, resulting in thousands of FECs. Therefore, plain MPLS in an OTN does not work because the label space that DWDM networks typically provide is 40 to 128 labels (wavelengths) per port.

As a consequence, MPLS-TE (with some hierarchical extensions) is used in the MPLmS architecture to reduce the number of required labels. The fundamental element within the MPLS-TE architecture is the ***traffic trunk***. A traffic trunk is an aggregation of traffic flows forwarded to the same destination and belonging to the same traffic class. A traffic trunk is represented by an LSP. Certain attributes can be assigned to a traffic trunk, such as bandwidth requirements, sensitivity, priority, preemption, resilience, and resource class affinity. When deploying the MPLS-TE architecture in an OTN, a lightpath is carrying several LSPs having the same destination and compatible LSP attributes. Thus, the term ***traffic trunk*** can be associated with the term ***lightpath***.

## MPLmS Model

MPLmS makes it possible to have one single control plane for both the optical transport layer and the service layer. This uniform control plane is based on MPLS and collapses both the optical transport and the service layer into a single layer.

In this architectural model, the IP routers of the service layer and the WRs of the optical transport layer maintain peer relationships between each other (Figure 4–42). Because all network elements are part of the same routing domain, the topology of the OTN is visible to the service layer network and vice versa.

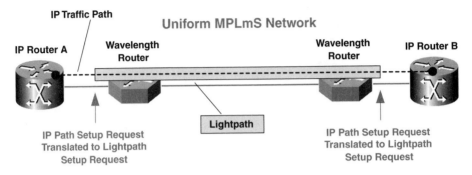

**IP Traffic Path**

**Uniform MPLmS Network**

IP Router A

Wavelength
Router

Wavelength
Router

IP Router B

Lightpath

IP Path Setup Request
Translated to Lightpath
Setup Request

IP Path Setup Request
Translated to Lightpath
Setup Request

**Figure 4–42**   In the peer model, both IP routers and WRs are part of the same routing domain and act as peers

Connections can be provisioned across the network transparently without any manual interaction or translation at the edge of the OTN. LSP setup requests are seamlessly transformed between LSRs and WRs into lightpath setup requests as both similar MPLS mechanisms are used.

## Architecture and Elements

According to the architectural model described earlier, an MPLmS network consists of WRs and is surrounded by LSRs. A typical MPLmS network is shown in Figure 4–43.

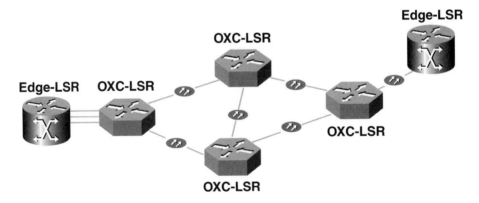

Edge-LSR

OXC-LSR

Edge-LSR   OXC-LSR

OXC-LSR

OXC-LSR

**Figure 4–43**   OTN with WRs and LSRs

These LSRs at the edge are also called ***Edge-LSR***s and have two functions. First, traffic flows of the service layer network are aggregated to high-bandwidth traffic streams, suitable for efficient use of the limited number of available light-paths. Second, Edge-LSRs request unidirectional lightpaths (also called LSPs) to be set up by the WRs through the OTN.

WRs with MPLmS control planes are referred to as ***OXC-LSRs***, analogous to ATM and Frame Relay (FR) switches, called ***ATM-LSRs*** and ***FR-LSRs***.

## MPLmS Control Plane

The generic requirements to the OXC control plane are to provided the capability to establish optical channels, to support traffic engineering functions, and to offer protection and restoration mechanisms.

Different link properties must be taken into account when establishing a lightpath through the OTN. The performance of a lightpath depends on the characteristics of the used links. Therefore, there must be a possibility to restrict the path to be taken by the lightpath to ensure requested performance characteristics. Mechanisms must activate and deactivate lightpaths, verify the proper operation, and assign them to LSP. This includes providing details to the OXC to reconfigure the cross-connect table of the OXC. Of course, spare links for protection purposes also have to be identified.

The functional building blocks of the MPLmS control plane, as shown in Figure 4–44, are similar to the standard MPLS-TE control plane. A link-state Interior Gateway Protocol (IGP), which can either be OSPF or IS-IS with optical domain-specific extensions, is responsible for distributing information about OTN topology, resource availability, and network status. This information is then stored in the traffic engineering database. A constraint-based routing function acting as a path selector is used to compute routes for the LSPs through the mesh optical network. Signaling protocols such as RSVP-TE and CR-LDP are then used to set up and maintain the LSPs by consulting the path selector.

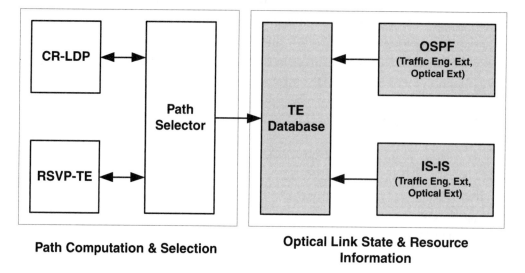

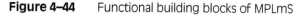

**Path Computation & Selection**    **Optical Link State & Resource Information**

**Figure 4–44**   Functional building blocks of MPLmS

An important fact to note is that the lightpath or LSPs are dynamically switched through the OTN but are terminated at the Edge-LSRs.

The MPLmS control plane for OXCs cannot fulfill two functions.

First, the analog to label merging does not exist. An OXC-LSR cannot merge traffic coming from two different LSPs into one single LSP at the optical domain because the LSP is represented by a wavelength and not by a label within a packet header.

Second, an OXC-LSR cannot perform an equivalent to the label push and POP functions of service layer LSRs. Furthermore, OXC-LSRs with integrated DWDM terminals maintain a Lambda Forwarding Information Base (LmFIB) per trunk, because the same channel can be assigned at each span to which the WR is connected. Comparing this with the restrictions of most ATM-LSR implementations, which do not support VC-merge and label push/pop, a WR using MPLmS is similar to an ATM-LSR.

Similar to the wavelength routing control plane of the overlay model described in the previous section, the MPLmS control plane requires a DCN running IP to exchange control information. As described in the previous section, fixed configured lightpaths, interface framing overhead, or a separate infrastructure can be used. When implementing a MPLmS control plane for OTNs, several issues have to be considered.

First, it must be defined what type of OTN state information has to be distributed by the IGP. This might also include optical domain-specific characteristics encoded as metrics, such as optical attenuation and dispersion. Furthermore, the constraints used for computing LSPs have to be defined.

Second, the OXCs must be enhanced with MPLmS functionality. Again, basically, there are two ways of implementing the MPLmS control.

The first one is to attach an external router running MPLmS via a standard control interface to each OXC. These routers may be referred **as *Lambda Signaling Controllers*** (LmSCs) and provide functions such as optical resources management, configuration and capacity management, addressing, routing, traffic engineering, topology discovery, and, of course, restoration. Again, the control interface specifies a set of primitives such as "connect," "disconnect," or "switch" used by the LmSC to configure the OXC. An OXC vendor-specific device translates between the standard control interface primitives and the proprietary controls of the OXC.

The second way is to integrate MPLmS functionality into the OXC. Both approaches are shown in Figure 4–45.

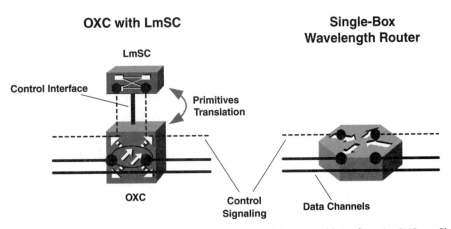

**Figure 4–45**    Lambda Switch Controller (LmSC) with control interface to OXC vs. Single-Box WR

MPLmS is completely independent of underlying OXC implementation. An MPLS control plane has already been implemented on IP routers (IP-LSRs), ATM-LSRs, and FR-LSRs. Therefore, as shown in Figure 4–46, it is obvious that MPLS can also be taken for an OXC control plane by using an adaptation function that maps MPLS functionality onto the OXC-specific controls.

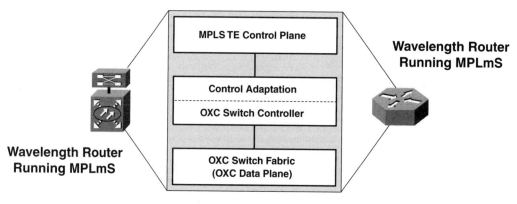

**Figure 4–46**    MPLmS control plane Architecture

## Lightpath Provisioning

### Optical Routing

OXC-LSRs and Edge-LSRs are running an IGP, which can be either OSPF or IS-IS. The IGP is used to determine connectivity and to collect resource information necessary for the Edge-LSRs to compute the paths for requested LSPs. Information exchange is done by OSPF via flooding link state advertisements (LSAs) and by IS-IS via Link State Protocol data units. To avoid confusion with the abbreviation for label-switched paths (LSPs), both OSPF and IS-IS advertisements will be called *LSA* in this work.

The IGP domain includes all OXC-LSRs and Edge-LSRs. It is important to note that the Edge-LSRs do not redistribute any IGP routing information into IGP routing protocol instances of the service layer network. The OTN is a completely separated IGP routing domain because other routers behind the Edge-LSRs cannot process the extended IGP information. Only the established LSPs through the OTN are advertised as links into the IGP of the service layer network.

Extensions to OSPF and IS-IS for optical routing are still under study. There have already been proposed some drafts to the IETF. The referenced IETF drafts provide the basis of the following paragraphs.

## TOPOLOGY DISCOVERY

It is important for all network nodes, including the OXC-LSRs and Edge-LSRs, to know the whole topology of the network. Before Edge-LSRs and OXC-LSRs can exchange topology information, neighbor link discovery has to occur. After the link discovery process has been finished, each node knows which channel of a local port is connected to which channel of a remote port of the neighboring node.

Instead of a link discovery protocol, manual configuration could also be considered. However, the more scalable and preferable way is to use a kind of Hello protocol to discover neighbors and channel interconnectivity information.

Hello packets are sent out periodically, one on each output port on each channel containing an LSR-ID and the port/channel combination. As LSR-ID, the IP address of a logical loopback interface might be used. The LSR on the other side receives these packets if the channels of the input ports are compatible. Each channel where Hello packets are periodically received is included in the port state database. Figure 4–47 shows an example for such a port state database maintained by processing Hello packets.

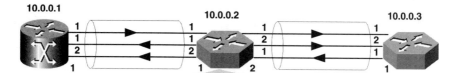

| Port State Database for 10.0.0.2 | | | | |
|---|---|---|---|---|
| **Local Output Port** | | **Remote Input Port** | | |
| Port # | Channel # | Node ID | Port # | Channel # |
| 1 | 1 | 10.0.0.1 | 1 | 1 |
| 1 | 2 | 10.0.0.1 | 1 | 2 |
| 2 | 1 | 10.0.0.3 | 1 | 1 |
| 2 | 2 | 10.0.0.3 | 1 | 2 |

| **Local Input Port** | | **Remote Output Port** | | |
|---|---|---|---|---|
| Port # | Channel # | Node ID | Port # | Channel # |
| 1 | 1 | 10.0.0.1 | 1 | 1 |
| 2 | 1 | 10.0.0.3 | 1 | 1 |

**Figure 4–47**   Port state database

The port state database includes information about all active channels between neighboring nodes; thus, it represents the available resources.

### OPTICAL LINK STATE ADVERTISEMENT

Control channels are assigned between the OXC-LSRs and Edge-LSRs to establish neighbor relationships for the IGP. The LSRs are interconnected via multiple links provided by DWDM terminals or multiple fibers. All links with the same traffic engineering characteristics between two LSRs are grouped together. They will have a *logical control channel* assigned and form an *IP link*. Thus, if there are links with different traffic engineering characteristics between two LSRs, multiple IP links are to be created, each with its own control channel. A common physical control channel of the DCN is used for setting up multiple logical control channels of the IP links.

OSPF or IS-IS uses the logical control channels to establish neighbor relationships across the IP links and to carry low-bandwidth control traffic. This includes flooded LSAs of the IGP, RSVP messages, telnet sessions, and so on.

To control the number of LSAs to be flooded and to avoid link-state flapping, thresholds can be defined. Thus, an LSA for a link is triggered only when the difference between the last advertised bandwidth and the current available bandwidth crosses the threshold of that certain link.

### OPTICAL LSA OBJECTS

OSPF and IS-IS have already been extended to support traffic engineering functions in IP networks. For OSPF, opaque LSAs and, for IS-IS, new TLVs have been defined. The way to the extensions for optical routing is straightforward. Already, defined traffic engineering extensions are used as a basis and are complemented with some additional Type/Length/Values (TLVs) to be able to transport information, such as link types and number of available channels within the IGP.

For details on OSPF and IS-IS traffic engineering extensions, refer the MPLS-TE section in Chapter 3 or the two IETF drafts, "Traffic Engineering Extensions to OSPF" [IETF-17] and "IS-IS extensions for Traffic Engineering" [IETF-23].

For both OSPF and IS-IS, the LSA payload contains multiple top-level Type/Length/Value (TLV) objects. In OSPF, they are called the *Router Address TLV* and the *Link TLV*. Within IS-IS, they are called *Router ID TLV* and *Extended Reachability TLV*.

These top-level TLVs include some nested sub-TLVs. Each TLV or sub-TLV has the format shown in Figure 4–48.

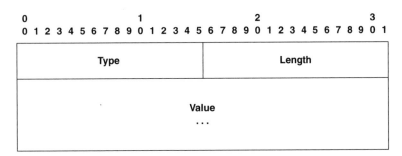

**Figure 4–48**    TLV and sub-TLV format

The Type field encodes the type of TLV or sub-TLV. The Length field specifies the length of the value field in bytes. The Value field contains various parameters, such as link type, bit rate, or encoding scheme describing the optical network resources, and may also contain an arbitrary number of sub-TLVs.

The IETF drafts "OSPF Extensions in Support of MPL(ambda)S" [IETF-15], "IS-IS Extensions in Support of MPL(ambda)S" [IETF-16], "Extensions to IS-IS/OSPF and RSVP in Support of MPL(ambda)S" [IETF-18], and "Extensions to CR-LDP and RSVP-TE for Optical Path Setup" [IETF-20] propose various IGP extensions and define changes for the Link TLV of OSPF and Extended IS Reachability TLV of IS-IS.

The goal is to define standardized extensions for the already existing sub-TLVs and a set of new sub-TLVs, to be able to distribute parameters, such as protection characteristics, bit rate, and encoding of the several links, between the network elements. These parameters define certain link types necessary for constrained-based routing done by the Edge-LSRs. Detailed descriptions of sub-TLV formats are out of the scope of this work because the exact definitions are still under investigation, but there are several considerations to be covered.

Each fiber span between OXC-LSRs and Edge-LSRs provides different types of links, depending on the physical OXC-LSR and Edge-LSR interfaces and the capability of the DWDM terminals. Depending on the *multiplexing capability* of the OXC-LSR and Edge-LSR interfaces, data received on a link might be processed at the packet, TDM, or wavelength level.

Each link may support various *bit-encoding formats* and *transmission rates,* which must also be considered during traffic engineering. For instance, a link may support a SONET or SDH signal up to 2.5 Gbps (OC-48/STM-16) and a 1-Gbps Ethernet signal but not a 10-Gbps (OC-192/STM-64) signal. Another example is a TDM link supporting up to 16 STM-1 signals, up to 4 STM-4 signals, or a single STM-16 signal, depending on the current resource availability. A DWDM link may support several STM-192 or 10-Gbps Ethernet signals through DWDM.

OXC-LSRs and Edge-LSRs may be interconnected with DWDM equipment providing protection at the optical layer. Hence, the DWDM channel of the link may be protected by another prereserved and physical path disjointed DWDM channel. The type of protection provided by the underlying layer (DWDM or also SONET/SDH) is indicated by the *protection type* of a link.

A network failure, such as a fiber cut or DWDM terminal breakdown, may affect several links at a time. These links may be assigned to a *risk group*, which is then used as a metric for traffic engineering. Because a link may be affected by several failures, it is assigned to more than one of these risks. Each of these risk groups is unique within the IGP domain of the OTN. The risk group of a whole end-to-end path is the union of the risk groups of all used links. When computing diversely routed paths for LSPs, path computation must find paths with disjointed risk groups.

A set of sub-TLVs has to be defined to carry all this characteristic information, which is then taken into account when computing LSP paths through the optical network. For example, the bit rate of all links used by an LSP across the network must be the same and equal to at least the desired LSP's bandwidth. Of course, the encoding must also be the same.

Already defined sub-TLVs, such as the Resource Class sub-TLV, may also be applied. This sub-TLV carries a 32-bit value used by the constrained-based routing mechanism to distinguish whether to include or exclude a certain link into the path for an LSP through the OTN.

### Constrained-Based Routing

The constrained-based routing algorithm resides on top of the standard IGP routing process. It is responsible for calculating explicit paths for LSPs satisfying the demand requirements.

The Constrained Shortest Path First (CSPF) algorithm first analyzes available resources, then compares them with the requested attributes of the LSP.

Taking into account that some links and nodes in the OTN do not satisfy the requirements, it runs an SPF calculation and chooses the path.

The constrained-based routing algorithm provides the explicit route to the signaling protocol, which is then used for LSP setup.

## Lightpath (Optical LSP) Signaling

As mentioned earlier, LSP setup within the OTN is signaled using RSVP-TE or CR-LDP. RSVP-TE and CR-LDP are extended versions of RSVP and LDP to support traffic engineering functionality.

The IETF draft "Signaling Framework for Automated Provisioning and Restoration of Paths in Optical Mesh Networks" [IETF-19] describes general issues to be taken into account for LSP setup in an OTN. The IETF draft "Extensions to CR-LDP and RSVP-TE for Optical Path Setup" [IETF-20] proposes very detailed extensions to both RSVP-TE and CR-LDP and describes optical signaling operations. Considering the use of MPLS not only in IP or ATM but also in optical networks, MPLS signaling has to be generalized. The required extensions or changes to RSVP and LDP are proposed in the IETF draft "Generalized MPLS—Signaling Functional Description" [IETF-21].

The following paragraphs are based on these proposals and describe the LSP setup procedures for MPLmS in a more general notion.

### CONCEPT OF NESTED LSPs

Network operators are moving toward traffic engineered IP networks. An obvious approach is to take the service layer network and integrate it into the OTN. The concept of MPLmS assumes that the routers interfacing the OTN, referred to as Edge-LSRs, will aggregate LSPs of the service layer sourced and terminated at the common nodes, which are then tunneled over an LSP set up through the OTN. In other words, service layer LSPs are nested into optical LSPs.

As proposed in the IETF draft "LSP Hierarchy with MPLS-TE" [IETF-22], LSPs through the OTN, which are nesting LSPs of the service layer, are called *forwarding adjacency* (FA) and are advertised as links into the link-state database of the IGP. There are several types of FAs, depending on what link type should be supported. Thus, there is an FA for LSPs, where data is processed at the packet level, another one for TDM signals, and so on.

FAs act as links used by other LSPs and are advertised into the IGP. For example, during path computation, it is important for the IGP to know what path is followed by the LSP acting as FA. Therefore, the LSAs advertising an FA also include the explicit hop-by-hop path in a certain TLV object.

FAs are virtual one-hop shortcut links through the OTN and, therefore, must not be used for establishing IGP peering relations. FAs are used only to determine the optimum path through the OTN by the CSPF algorithm.

Consider the situation where MPLmS is used to implement an overlay model. In this case, there are two separate MPLS-TE planes for the service and optical transport layer. FAs are created and maintained by the MPLS-TE plane of the OTN and provided to the service layer MPLS-TE control plane to be used as links by the constrained-based routing mechanism.

### Optical LSP Setup

By using MPLS traffic engineering for the control plane of the OTN, the LSP setup is controlled by the head-end Edge-LSR. The Edge-LSR knows the whole network topology and current resource allocations through its IGP process. It uses the CSPF algorithm to compute the best path through the OTN and performs source routing by defining an explicit path that must be used during the LSP setup.

LSPs can be set up either statically by a Network Operation Center (NOC) performing off-line network optimization and traffic engineering or dynamically by the Edge-LSRs, which are taking part in the service layer IP routing process.

There has already been a long and intensive discussion between different router vendors as to whether to use LDP or RSVP for LSP setup in MPLS. Some decided to extend LDP, and some decided to extend RSVP. Because the trend seems to move toward RSVP-TE, the following descriptions of LSP setup procedures concentrate on RSVP-TE.

### *RSVP SETUP PROCEDURE*

To set up an LSP through the OTN, the head-end Edge-LSR is sending a `PATH` message including a `LABEL_REQUEST` object and an `EXPLICIT_ROUTE` object (ERO) toward the downstream next hop. The `LABEL_REQUEST` object indicates a channel (label) allocation request for a certain link type and provides a pointer to the upper layer protocol to be transported over the LSP. Referring to Figure 4–49, each next-hop OXC-LSR performs a bandwidth availability and link type compatibility check. If both checks are passed, the OXC-LSR removes the first subobject of the ERO—indicating itself as next hop—and sends a `PATH` message including the modified ERO toward the next hop. If the OXC-LSR is unable to support the desired connection, it generates a PATHERR message and sends it toward the head end of the LSP.

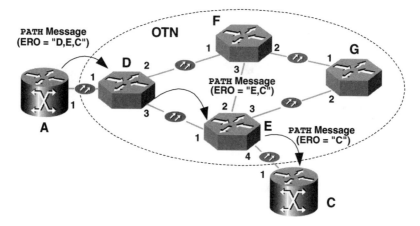

**Figure 4–49**   The head end initiates the optical LSP setup by sending an RSVP PATH message

As can be seen in Figure 4–50, The Edge-LSR terminating the LSP is responding to the PATH message by sending a RESV message towards the head-end Edge-LSR. Within the RESV message, the LABEL object is included to inform the upstream OXC-LSR which channel is to be used on the link between itself and the upstream OXC-LSR. The upstream OXC-LSR receives the RESV message and extracts the LABEL object to determine which channel to use as output for that LSP. To get the corresponding input, the OXC-LSR looks up its port state database and determines the channel providing connectivity to the next upstream OXC-LSR of the LSP path. Having the information about both the input and output channel, the OXC-LSR can update its cross-connect configuration.

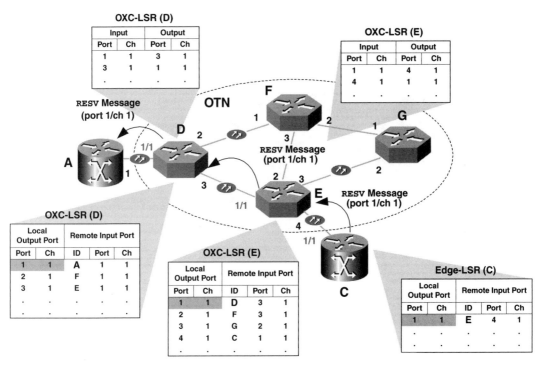

**Figure 4–50**    The tail end finishes the optical LSP setup by responding with an RSVP RESV message

This procedure is completed by each hop along the path. After the RESV message is received by the head-end Edge-LSR, the LSP is set up completely.

### OPTICAL LSP SETUP IN AN OTN WITHOUT WAVELENGTH CONVERSION

Some OXC-LSRs may be wavelength routers with restricted or no wavelength conversion capability. An example is if the OTN running MPLmS includes a transparent optical subnetwork. In such a situation (shown in Figure 4–51), PATH messages may contain a LAMBDA_SET object. The LAMBDA_SET object, sent by the Edge-LSR, contains all the wavelengths it is able to assign. The downstream OXC-LSR in the path checks this LAMBDA_SET object against the wavelengths it is able to allocate, removes all already assigned wavelengths, and sends this modified LAMBDA_SET object to the next OXC-LSR. At the end of the path, the Edge-LSR also checks the LAMBDA_SET object, selects one of the remaining wavelengths, and includes this wavelength in the LABEL object of the RESV message.

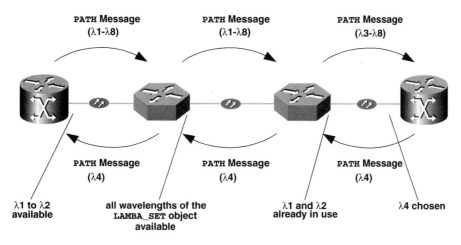

PATH Message
($\lambda$1-$\lambda$8)

PATH Message
($\lambda$1-$\lambda$8)

PATH Message
($\lambda$3-$\lambda$8)

PATH Message
($\lambda$4)

PATH Message
($\lambda$4)

PATH Message
($\lambda$4)

$\lambda$1 to $\lambda$2
available

all wavelengths of the
LAMBA_SET object
available

$\lambda$1 and $\lambda$2
already in use

$\lambda$4 chosen

**Figure 4–51**    LSP setup in an OTN without wavelength conversion

*LSP AGGREGATION*

As already mentioned and shown in Figure 4–52, LSP aggregation is done at the edge of the OTN. Supposing RSVP as the signaling protocol, Edge-LSRs are receiving PATH messages for service layer LSPs to be transported over the OTN. By processing the ERO of the PATH message, an Edge-LSR verifies that it is the current hop for the LSP, and it determines which of the other Edge-LSRs of the OTN is the next hop for the LSP. Then the CSPF algorithm computes the optimum path. Next, the Edge-LSR looks for an already established FA with the appropriate link type parameters and matching path computed by the CSPF algorithm. If there is such an FA with enough unreserved bandwidth available, the Edge-LSR uses it. Otherwise, a new FA has to be established. This is done by going through the optical LSP setup procedure, described earlier.

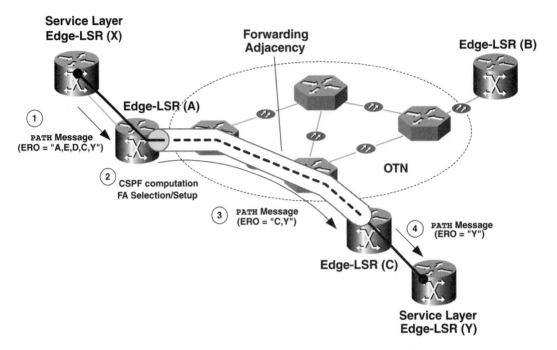

**Figure 4–52**    LSP aggregation at the OTN edge, using RSVP signaling

The Edge-LSR removes itself, and the hops part of the FA from the hop list of the ERO and sends it in a PATH message to the other Edge-LSR. The Edge-LSR at the terminating end of the FA also processes the ERO and forwards the PATH message to the next hop.

Finally, the node at the endpoint of the aggregated LSP sends a RESV message back along the path to initiate bandwidth reservation. For the FA of the OTN, the head-end Edge-LSR allocates the required bandwidth for the LSP by simply subtracting the bandwidth of the LSP from the unreserved bandwidth of the FA.

## Using MPLmS in the Overlay Model

The advantage of MPLmS is that the optical transport and service layer can be collapsed into one layer through deployment of a uniform control plane. This approach is obvious from the technical point of view but might not be obvious from an administrative point of view.

Some service providers do not want to build an IP infrastructure and sell IP services. For example, interexchange carriers might want to concentrate on selling high-bandwidth connectivity through their OTN. To optimize and simplify their optical network, they want to use wavelength routing. To use a standard solution, they might want to use MPLmS and not a proprietary implementation. In this case, the control plane of the interexchange carrier's optical network and the IP network of its customers operate independently. Customers use MPLS in their IP network, but interworking between the control planes is done via static configuration or some dynamic procedures.

## Restoration

Restoration in MPLmS networks is similar to wavelength routing networks, as described in a previous section. The main differences are that, first of all, standard MPLS mechanisms and not proprietary protection mechanisms are used. Typically, 1:1 or 1:n protection incorporated through the MPLS-TE functionality is applied.

As described in Chapter 3, "Optical Networking Technology Fundamentals," link or node protection might be applied. In doing this, optical LSPs (O-LSPs) are predefined and preestablished around nodes or links likely to fail. In case of a failure, no end-to-end signaling is required to restore the traffic affected by the failure. The nodes adjacent to the failure simply switch all nested LSPs onto the backup OLSP.

To optimize the network utilization, path protection is used. Using path protection requires signaling between the endpoints of an OLSP in case of a failure. This typically results in a longer restoration time.

## Summary

This chapter guided us through the three evolutionary steps, from traditional OTNs to IP-optimized, next-generation OTNs. The first step included the elimination of unnecessary transmission overhead and network complexity by eliminating both the ATM and SONET/SDH network layers. IP routers directly connected to DWDM systems provide up to terabit of capacity through static-provisioned connections; therefore, we call this solution the *Static IP Optical Overlay Control Plane*.

The second step addresses the inefficiency of statically provisioning optical connections between IP routers. The optical core network is comprised of a mesh of DWDM systems interconnected by WRs. These WRs facilitate dynamic provisioning for connections through the transport network core by running an optical routing protocol. Enhanced optical mesh protection is also possible through the rerouting of failed optical connections.

By using on-demand mesh protection instead of ring protection with preserved bandwidth, the overall transmission capacity of the transport network can be optimized. Because this dynamic optical transport network is opaque to the overlaying IP network above, this step is also referred to as the **_Dynamic IP Optical Overlay Control Plane_**.

The third step in the evolution of OTNs highly focuses on the integration of the optical transport layer and the IP layer. A uniform control plane for both the optical transport and the IP layer is to be standardized in order to allow the OTN to be visible to the IP network and to integrate the provisioning processes of both layers. One approach for this uniform control plane is MPLmS. MPLmS basically adapts the MPLS-TE architecture concepts to the optical transport domain. This approach allows network operators owning both the optical transport and the IP part of the network to create a network according to the **_Integrated IP Optical Peer Control Plane_** model.

# Recommended Reading

## Static IP Optical Overlay Control Plane

[AF-1] ATM Forum Specification, af-lane-0021.000, _LAN Emulation over ATM Version 1.0_, January 1995.

[CSCO-3] Cisco Systems Inc. Whitepaper, _Dynamic Packet Transport Technology and Applications Overview_, January 1999.

[CSCO-6] Cisco Systems Inc. Whitepaper, _Designing ATM MPLS Networks_, Jeremy Lawrence, April 1999.

[CSCO-7] Cisco Systems Inc. Whitepaper, _ATM Internetworking_, March 1995.

[CSCO-8] Cisco Systems Inc. Whitepaper, _Building High-Speed Exchange Points_, Revision 1.2, Paul Ferguson, March 1997.

[IETF-6] IETF-draft, draft-ietf-mpls-arch-02.txt, *Multiprotocol Label Switching Architecture*, work in progress, July 1998.

[IETF-7] IETF-draft, draft-ietf-mpls-traffic-eng-00.txt, *Requirements for Traffic Engineering over MPLS*, work in progress, October 1998.

[JUNP-1] Juniper Networks Whitepaper, *Traffic Engineering for the New Public Network*, January 1999.

[MINO-1] McGraw-Hill Companies, *IP Applications with ATM*, J.Amoss, Ph.D., D. Minoli, May 1998.

[NFOEC-1] Technical Paper, *"IP over WDM" the Missing Link*, P. Bonenfant, A. Rodrigues-Moral, J. Manchester, Lucent Technologies, A. McGuire, BT Laboratories, September 1999.

[TELL-1] Tellium Whitepaper, *WDM Optical Network Architectures for a Data-Centric Environment*, Krishna Bala.

## Dynamic IP Optical Overlay Control Plane

[IETF-11] IETF-draft, draft-ip-optical-framework-00.txt, *IP over Optical Networks—A Framework*, work in progress, March 2000.

[MUKH-1] McGraw-Hill, *Optical Communication Networks*, Biswanath Mukherjee, 1997.

## Integrated IP Optical Peer Control Plane

[IETF-11] IETF-draft, draft-ip-optical-framework-00.txt, *IP over Optical Networks—A Framework*, work in progress, March 2000.

[IETF-13] IETF-draft, draft-awduche-mpls-te-optical-01.txt, *Multi-Protocol Lambda Switching: Combining MPLS Traffic Engineering Control with Optical Crossconnects*, work in progress, November 1999.

[IETF-14] IETF-draft, draft-basak-mpls-oxc-issues-01.txt, *Multi-Protocol Lambda Switching: Issues in Combining MPLS Traffic Engineering Control with Optical Crossconnects*, work in progress, November 1999.

[IETF-15] IETF-draft, draft-kompella-ospf-ompls-extensions-00.txt, *OSPF Extensions in Support of MPL(ambda)S*, work in progress, July 2000.

[IETF-16] IETF-draft, draft-kompella-isis-ompls-extensions-00.txt, *IS-IS Extensions in Support of MPL(ambda)S*, work in progress, July 2000.

[IETF-21] IETF-draft, draft-ietf-mpls-generalized-signaling-00.txt, *Generalized MPLS—Signaling Functional Description*, work in progress, October 2000.

[IETF-22] IETF-draft, draft-ietf-mpls-lsp-hierarchy-00.txt, *LSP Hierarchy with MPLS TE*, work in progress, July 2000.

[MUKH-1] McGraw-Hill, *Optical Communication Networks*, Biswanath Mukherjee, 1997.

# 5

# Optical Networking Applications and Case Examples

## Optical End-to-End Networking Design Trends

It should be very obvious that there is no single optimal design for all the different kinds of optical networks that can be applied to cover all requirements, as well as handle all applications. The network architecture must be adapted to the different needs and the existing constraints. This is further complicated because there are big differences in the cases where optical networks can be deployed.

Therefore, we will attempt to outline in this chapter the different needs for the most important deployment scenarios of optical networks we can define today, these being the metro network, the service point of presence (service POPs), and the core network. We do this in order to provide the best design to support the offer and guarantee of the services and applications in these areas.

Let us first provide a snapshot of the way we see the Internet growing, including the types of traffic that are coming into the network and how this traffic is being passed. We will then use this as a reference for several different concept introductions for the different optical network deployment scenarios.

As we are probably all aware, the Internet is changing the foundations of our lives. What we find is that, as this situation reaches critical mass, people are expecting things from the Internet, and we have to deploy the right solutions for the different scenarios to make these things become a reality.

In most cases, if service providers and enterprises are faced with issues in their networks, they are looking for help with scaling their solutions and providing more and more network capacity. However, it is not about just scaling the network architecture; it is also about providing services at a reasonable cost. This means that IP + optical solutions have to be designed with the capital and operational expenses (CAPEX and OPEX) in mind so that they offer solutions that are truly future-proof. This will put the implementers in the position to be ahead of the competition.

Whether those services are capacity-oriented, content-oriented, or application-oriented, the real goal is to give service providers the ability to differentiate themselves and offer rapid service creation and delivery, quick time to market, and the ability to meet the needs of emerging markets.

One thing that all these applications share in common is the Internet Protocol (IP). Lacking the ability to understand and process information at the IP layer only allows the offering of simple services. As technology and requirements evolve, we find that the Internet business models are strengthening the need for IP, not eliminating it. We all remember analyst reports saying that with ubiquitous end-to-end optical networking, the need for network layer routing will simply disappear. Today, we are seeing a completely different effect—the need for intelligent end-to-end control of the optical networking plane in all application areas; metro, POP, and core will not go away. This control is needed now more than ever, and it must be based on IP!

## Service POP

Initially, when we started building pre-Internet networks, most service requests were satisfied within the building, or within the campus. Only about 20% of the time did requests actually leave the campus—for things like e-mail or file transfer (see Figure 5–1).

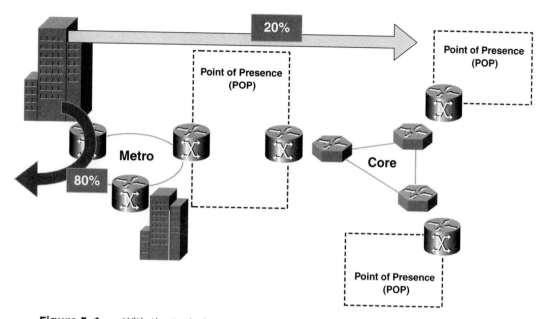

**Figure 5–1**   With the typical pre-Internet LAN/WAN traffic patterns, 80% of the traffic stayed within the building or campus network and only 20% left toward other locations

Over the past few years, we have seen the Internet emerge dramatically, meaning that we have simply learned to deal with Internet traffic patterns. What we see today are traffic patterns completely similar to the original Local Area Network (LAN)/Wide Area Network (WAN) patterns, well known from the pre-Internet networks. Now, however, 80% of the time, the traffic leaves the enterprise and stays within metro areas, and only 20% of the time does that traffic move between major metro areas across the core (see Figure 5–2). This has a lot to do with the transformation of the points of presence (POPs), which we will discuss next.

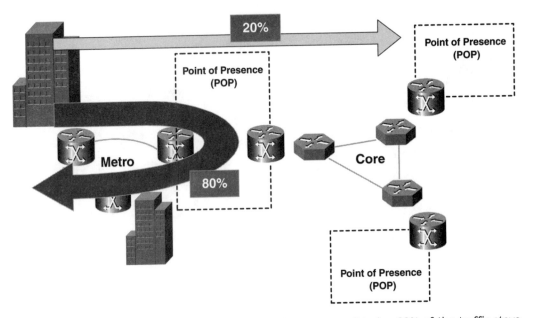

**Figure 5–2**   With the changed Internet traffic patterns of today, 80% of the traffic stays within the metro area, and only 20% leaves toward the network core

POPs acting as interconnects were, in fact, historically seen as central offices. These interconnects were not extremely intelligent, just fast and easy to manage at the management interface. Now, with the emergence of the Internet and data traffic, we have seen a big change here. The importance of the simple "old-world" point-to-point (p-t-p) interconnections is vanishing. Therefore, we are seeing a tremendous increase in intelligent services at the POP (nowadays called *Service Interconnect* or *Service POP*), which acts as the common connection point between metro networks and core networks (see Figure 5–3).

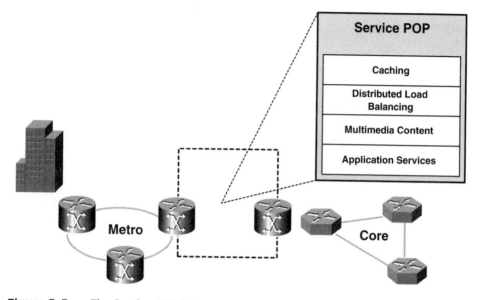

**Figure 5–3**  The Service POP interconnects the metro and core networks and delivers intelligent services to the user

So it makes sense to look a little bit deeper into the architecture of how these things are being built. On the metro side, the primary goal of service providers is to deliver customers to services. In the metro, if we believe we are delivering users to services, the user traffic leaves the building and is delivered to the Service POP in 90% of the cases. This delivery of customers to services must be as flexible as possible in order to aggregate users and move them between their desktops, i.e., their places of business, to where the services are delivered from the Service POP. What we end up with is a logical model where it makes sense to have a master node at the Service POP and client nodes at each of the customer sites. The main task of the master node is to offer connectivity to the core with all the additional services required here as storage, content, and caching services. At the client nodes, we typically see the customer premise connection, where we have to deal with caching, traffic classification, traffic prioritization, and traffic shaping in order to fulfill the Service Level Agreements (SLAs) negotiated between the customer and access service provider. Furthermore, we must not forget the Virtual Private Network (VPN) services provided in order to fulfill the privacy requirements of several customers.

In other words, it is all about aggregating traffic and taking the maximum advantage of the fiber plant to deliver users to services. This is actually the same concept that was implemented in the "old-world" central offices. The difference is that, today, we see many more advanced services optimized for Internet applications, e-commerce, business-to-business (B2B), and business-to-consumer (B2C) markets, compared with the relatively primitive services needed in the old-world economy.

On the other side of the Service POP in the core network, we have a slightly different set of challenges. Service providers recognize that the role of the core network is evolving quickly in order to support Internet traffic. We are dealing here with a tremendously established, huge market where we find Synchronous Optical Network/Synchronous Digital Hierarchy (SONET/SDH) rings, single mode fiber, 10-Gbps interfaces, and other well-established and well-known technologies—effectively 99% of the transport in core networks is handled and forwarded in that way. If we ask whether using these old-world technologies still makes economic sense today, we easily find out that it does not. In fact, later, when we talk about the core solutions, we will see why not.

## Metro Solutions

In the metro space, we have a fairly well defined set of challenges that can be addressed through optical networking. Once we get to the Service POP, we receive a different set of goals—we need to be IP- and application-aware. It is only through this IP + optical integration that we are going to be able to deliver content-rich and bandwidth-intense services.

Obviously, the biggest goal in delivering services is reducing the cost of per-user provisioning. The extent to which one can automate that process and bring user-to-network provisioning to a more efficient and automated state directly affects the ability to deliver the highest quality in radical economics, as well as profit margins.

So it is very important to reduce significantly per-user provisioning and to overcome the limitations of the last mile. We talk about last-mile limitations, where applications are going to drive multimedia content to the home—video delivery, distance learning, and remote access to high-speed shared applications. The market potential behind these applications is huge, as is the market for delivering high bandwidth to the home or to the business. We should also take into account that covering these areas with standards-based solutions is at least as important as reducing cost, provisioning time, and time to market, as well as overcoming bandwidth limitations.

## IP and Optical Metro Evolution

With this in mind, we see a logical evolution of the metro delivery space. We are literally entrenched today in a time division multiplexing (TDM)-based SONET/SDH era. Introducing *next-generation SONET/SDH* solutions brings us right to the heart of the next-generation metro network revolution, where companies can maximize their return of investment (ROI) on existing infrastructures, reducing the cost of deployment and increasing their productivity dramatically.

What next-generation SONET/SDH also means, for example, is that costs can be significantly reduced, because there is only one kind of transmission system left, instead of several types of multiplexers and cross-connects, as is shown in Figure 5–4. Service providers like this concept, because it enables them to connect customers for less, while providing services that differentiate. They can do this at a reasonable profit model, which finally guarantees their success.

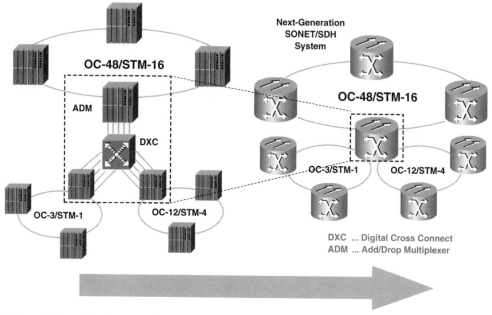

**Figure 5–4**    Next-generation SONET/SDH systems simplify the metro network architecture

What is next? One big issue is the role of a services optimization evolution, which means more intelligent use of bandwidth and more intelligent use of wavelengths with Dense Wave Division Multiplexing (DWDM) solutions for the metro space. Beyond increased gains in the next-generation SONET/SDH space, we see the evolution of more intelligent access technologies, such as Dynamic Packet Transport (DPT) to deliver IP-aware, essentially routed solutions over fiber to home and businesses. Therefore, the trend goes rather toward the deployment of one technology than toward having many in parallel.

Furthermore, we will see the evolution of Ethernet as a metro area connectivity tool. Enterprise customers love Ethernet, because it means a standardized way for easy provisioning. As soon as service providers will be able to offer Ethernet in the metro area as access technology, this will mean again a big differentiator and an important step to a further improved business model.

Overall, putting IP + optical together to deliver even more efficiency is the key. Many people are talking today about "IP over glass" and about merging IP and optical. *IP + optical* means, effectively, the integration of IP and optical, which we have described in the previous chapters as a way of controlling the optical plane by an IP-derived intelligence, or integrating IP routing and traffic engineering with the optical layer. It is effectively IP + optical that we are looking for, because building two separate infrastructures for data and transport in the coming years will not make sense any more. This is true not only for the metro access area; we will later see that very similar rules are also valid for the core of optical networks. Figure 5–5 shows the evolution of IP and optical in the metro area.

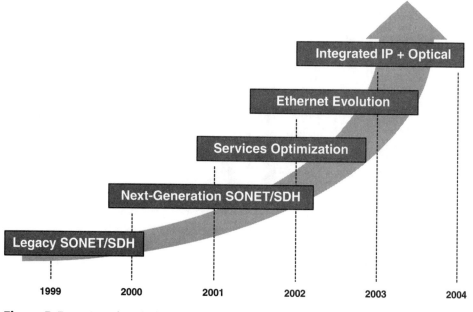

**Figure 5–5**    IP and optical metro evolution

It is important to mention here that there is no single optimal access method in the metro space. Being able to handle all of the different access types is critical, because it means that customers will have choices (see Figure 5–6). Being able to use a SONET or SDH infrastructure makes sense in many cases. Evolving an IP-based delivery mechanism, either based on Ethernet or based on DPT as well, makes a lot of sense in many cases. Taking advantage of already existing investment in IP + Asynchronous Transfer Mode (ATM) also makes sense. Therefore, it is important not to ignore any of these aggregation methods. Instead, it is important to understand how to get the most out of them by combining them in the most efficient way with the intelligent Service POP and enabling IP control of these infrastructures for optimal transport of today's IP-based applications.

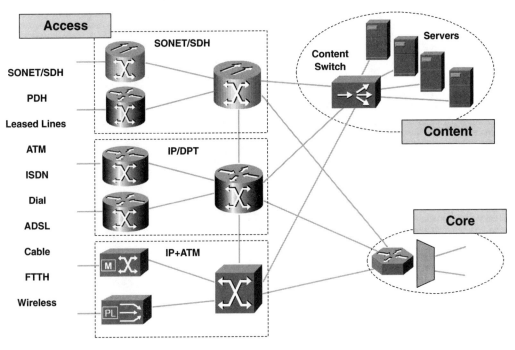

**Figure 5–6**   Different metro access methods

## Core Solutions

In the core, we find a slightly different set of challenges, as well as slightly different architectures. In the core, it is important to connect population centers and, in nearly all cases, to handle huge volumes of information. It is here where we want to make the most efficient use of the transport networks—switch wavelengths—while providing seamless high-performance connectivity between population centers and leave the electrical traffic processing to the edge. This is why the core network should be fundamentally traffic transparent, as opposed to the metro network, which is dominated by traffic grooming. In the Service POP, it is all about user service delivery, electrical processing, and software applications, and connecting to the edge of the optical, high-capacity core.

The challenge in the core is to meet today's and tomorrow's Internet demand. Internet traffic doubles every 100 days or even more. So here we have to deal with how to keep up with today's processing speeds and capacities, which is challenging. Fortunately, state-of-the-art technology gives us tools to handle this. Today's economy is more about how fast, not how much. Being able to turn up and adjust bandwidth more quickly in the core is key. This means that automated service creation is key in the transport market in the future, and those customers or those service providers that architect their infrastructure to take advantage of the evolution of service provisioning will soon outperform the rest.

## IP and Optical Core Evolution

Today, legacy SONET/SDH protection ring architectures at OC-192/STM-64 speeds are dominating. Recently deployed technologies, such as DWDM and the migration from ring structures toward peer-to-peer meshed environments, will gain even more importance over the next years.

Why is today's ring-based architecture fundamentally very wasteful? For a ring-based architecture to work and provide bandwidth from end to end across the country, the same amount of bandwidth has to be provisioned on every ring at every location. In the existing old-world environment, that has not been much of a problem as the traffic patterns have not changed a lot, and therefore, provisioning speed has not been that important.

Finally, now it is all about rapid end-to-end provisioning, and that's all about radical economics, which cannot be supported with old-world ring architectures. So, as was discussed in the previous chapters, there is a clear methodology to deploying meshed, intelligently controlled architectures (Figure 5–7).

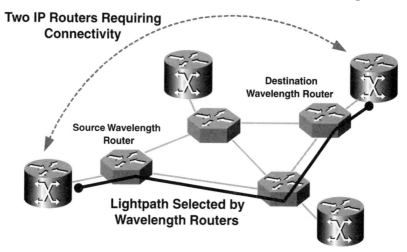

**Figure 5–7**   Migration to intelligently controlled meshed core architectures is key for service providers

Those service providers who adopt the transition from ring to mesh first will clearly be in a leadership position to deliver wavelength services, providing customers with an enhanced value proposition.

To take full advantage of these architectures, we are going to see the evolution of new control planes, which, again, will allow the integration of IP + optical as already discussed in the metro area. A standard way to do this is Multiprotocol Lambda Switching (MPLmS), as we have seen in the previous chapters—a natural extension of the work that was done in the ATM space, applied to the world of light, to deliver maximum potential from the advanced optical transport infrastructure. This integration of IP + optical will allow the core networks to grow and maximize their potential, very similar to the advantages this brings in the metro area. Figure 5–8 shows the IP and optical core evolution.

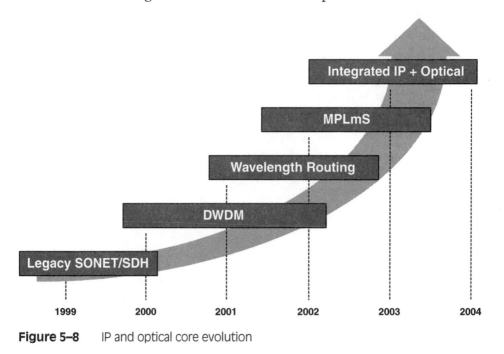

**Figure 5–8**    IP and optical core evolution

## Conclusions

As we have seen, scaling the Internet is obviously key. However, it is not just about scaling the Internet, it is also about raising service velocity. It is not just about lowering costs for service providers and enterprises, it is also about differentiated services. It is not just about moving bits, it is about moving packets intelligently. It is not just about incremental improvement, it is about radical economics.

If we look at the key challenges before us, the one thing that jumps right out at us is reducing operation, administration, maintenance, and provisioning (OAM&P) costs. OAM&P costs are still the maximum amount of costs for service providers—nearly half of the total costs—and also for enterprises, but maybe not to such a high extent that we see in the service provider space. This means that helping to reduce these costs is a key point for optical transport network (OTN) design and deployment.

As we come to the bottom line, IP + optical—as well as the combination of the intelligence of IP routing with the economics and capacity of optical technology—will allow the Internet of the future to evolve, grow, and be the foundation of intelligent service provisioning. This will occur independent of enterprise, metro, or core network areas. Changing Internet traffic patterns require advanced recognition of content and intelligent local distribution. Thus, the all-optical IP + optical network has a big future. We will demonstrate throughout the rest of this chapter that it makes a lot of sense to use optical technology to link major metro areas in order to take maximum advantage of the capacity and economics of optical. However, it also makes sense (because light has no intelligence and cannot be used as a computing medium today) to use IP to deliver those intelligent services.

## Case Example A: Next-Generation Storage Networks

In the past, enterprise data centers providing massive amounts of processing power usually used mainframes and were dominated by non-IP technologies. Mainframe vendors, such as IBM, developed their own set of protocols and network interfaces to be used to interconnect host computers and storage devices, such as tape systems or others.

IBM developed its Geographic Dispersed Parallel Sysplex (GDPS) solution. Figure 5–9 shows such a redundant GDPS mainframe configuration. There are two geographically separated data centers deployed. Each consists of a mainframe central processing unit (CPU), a Sysplex timer, a coupling facility (CF) and several storage devices, such as magnetic tape systems.

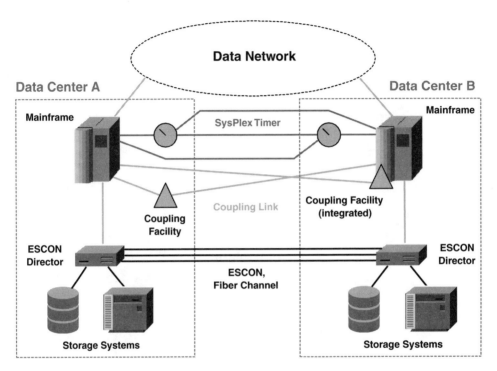

**Figure 5–9**    Enterprises typically have fully redundant data centers to ensure highest possible availability

This completely distributed system architecture is necessary because of the performance and high availability requirements.

The CPU has access to the storage devices through multiple *Enterprise Systems Connection* (ESCON) channels provided by the ESCON director, which is placed between the storage devices and the CPU. ESCON is an interface and protocol specification for transactions between CPUs and storage devices at speeds up to 200 Mbps. The ESCON director acts as a switch and controls which CPU has access to which storage device. Although ESCON is quite an old technology, 80% of all storage interconnections use ESCON. To allow a more efficient access to the storage devices, *Fiber Channel* (FC) has been developed. FC defines a 1.062-Gbps data connection used for the transport of protocols such as Small Computer System Interface (SCSI), which is typically used for communication with storage devices. Furthermore, it allows ESCON directors to bundle up to eight ESCON channels together into one FC.

During normal operation, both data centers are active. In case of a system failure, the nonaffected data center takes over all computation tasks and ensures that no running application is affected. To make this possible, the CPUs of both data centers must be exactly in sync, and the stored data must also be mirrored onto the storage systems of both data centers. The first requirement is fulfilled by the *Sysplex timer,* which is providing clocking for the CPUs. The second requirement is fulfilled by a 1.062-Gbps high-speed data connection called the *coupling kink* between both CPUs. The coupling link is terminated by the coupling facility, which might be integrated either into the CPU or an external system connected to the CPU. In addition, the storage systems are interconnected, using ESCON and FC, so that both data centers can have access to all storage systems available.

Detailed information about the GDPS system for IBM systems can be found in the IBM Redbook "OS/390 MVS Parallel Sysplex Configuration, Volume 1: Overview" [IBM-1].

This storage network has always been a completely separate island in the enterprise network, as shown in Figure 5–10. In this typical scenario, the network users are part of the IP-based LAN and are requesting stored information via the storage application servers that are acting as an interface to the attached storage network.

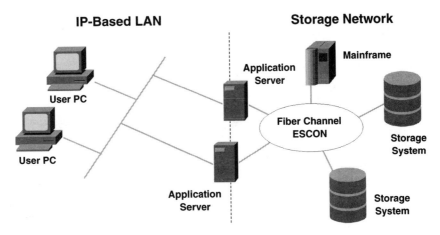

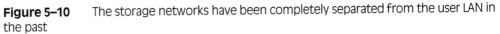

**Figure 5–10**    The storage networks have been completely separated from the user LAN in the past

As today's applications have evolved to become more global applications, the position of storage networking has also changed dramatically. Applications and stored information are no longer to be concentrated at a single, highly secure, and reliable place. Applications, storage, and processing power are distributed across the whole network. This trend calls for a completely new architecture, which is shown in Figure 5–11.

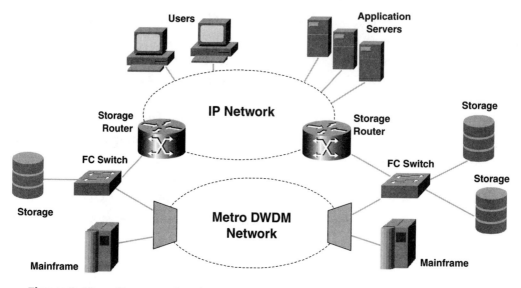

**Figure 5–11**    Storage networks are now to be integrated with the IP network to allow efficient information sharing

The first step shown on the bottom of Figure 5–11 is the introduction of DWDM technology to transparently transmit ESCON and FC across the network. This leads to the fact that storage networks are one of the most popular metro DWDM applications and an important driver for next-generation carrier networks.

The second step is shown in the upper part of Figure 5–11 and represents the integration of storage networking with IP technology. The Internet Engineering Task Force (IETF) has already developed a new protocol called *iSCSI* (SCSI transport over IP) to allow data exchange between storage devices and servers across an IP network. For detailed information, refer to the IETF draft, "iSCSI" [IETF-29].

In the following section, we want to concentrate on the use of metro DWDM for storage networking and discuss the key aspects of this kind of optical transport application.

## Application Requirements

The mainframe/storage scenario defines numerous requirements to the metro DWDM system. First, the DWDM systems preferably should have transparent transponders to be able to use the same transponder for ESCON, FC, and Sysplex for example. This opportunity reduces complexity and makes spare management easier. Metro DWDM systems available today typically provide transparent transponders in a bit rate range of 10 Mbps up to 2.5 Gbps, thus covering ESCON, FC, Sysplex, and other applications, such as ATM and Gigabit Ethernet. Because the coupling link is very sensitive to optical transmission impairments (including delay and signal quality), special transponders with 3R functionality at 1.062 Gbps are typically used.

Second, the number of signals that can be transported across the DWDM link must be in the range of 100–200. Typical GDPS installations require one or two channels for the Sysplex timer interconnects, at least one channel for the coupling link, then up to 100 or more channels for ESCON and FC storage interconnectivity. Metro DWDM systems mostly use the International Telecommunication Union (ITU) 200-GHz spacing and are, therefore, providing 32 wavelengths, at maximum, across one fiber. This is far away from the 200-channel requirement. Because of this, special TDM transponders are also used to multiplex, for example, four or eight ESCON channels onto one wavelength and to increase the channel capacity to the required level. This leads to a more efficient bandwidth use because one 800-Mbps or 1.6-Gbps TDM signal is transported across a wavelength, instead of one 200-Mbps ESCON signal.

The third and most important requirement is redundancy. The DWDM system must either be capable of performing optical protection switching to handle fiber cuts or it must be deployed in a complete redundant fashion.

## Common Solutions

Figure 5–12 shows such a typical GDPS application using a metro DWDM system. In this p-t-p application with the two data centers in two different locations, a fully redundant metro DWDM system is used. The various connections between the two data centers are shared across the two DWDM p-t-p paths. If one of the DWDM terminals fails, part of the connectivity between the data centers fails. This results in a decrease of the overall processing power of the GDPS solution but does not affect any running application.

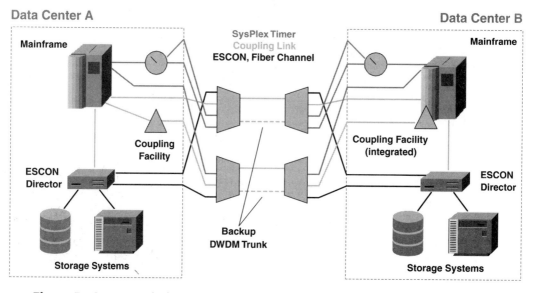

**Figure 5–12** Introducing a metro DWDM system with transparent transponders minimizes the amount of available fibers between the data centers

The highest level of redundancy is accomplished by also adding optical protection to the DWDM systems. As shown in Figure 5–12, two fiber trunks are used for both DWDM p-t-p paths. In case of a fiber cut, the loss of signal is detected at the receiving side, and optical protection restores the connection by switching on to the second fiber trunk. For details on optical 1 + 1 or 1:1 protection, refer to Chapter 4, "Existing and Future Optical Control Planes."

Some scenarios might require more than two data centers to be interconnected in a ring. Figure 5–13 shows such a scenario, where three data centers are interconnected ring-style. Because all storage connections, such as ESCON and Sysplex timer, are very sensitive to transmission impairments and delay differences, optical protection in a DWDM ring network can cause problems. It is an absolute requirement to guarantee that both the receive and the transmit paths of (for example) an ESCON connection are routed across the same side of the ring in any circumstance. Otherwise, the different path distances between the nodes across the ring will introduce a transmission delay difference and will make the ESCON connection fail.

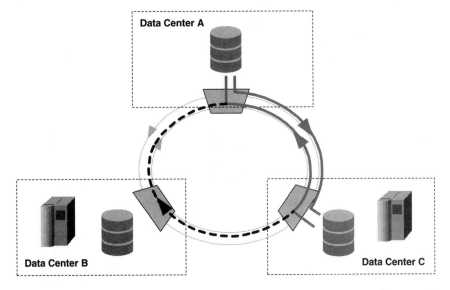

**Figure 5–13**  The receive and transmit paths of each connection, such as Sysplex timer, ESCON, and others in the ring, must be on the same side of the ring to avoid different transmission delays

## Case Example B: Next-Generation Internet Service Provider

The focus of an Internet service provider (ISP) is to deliver Internet access to its customers. These might be the broad mass of consumers, small and medium business customers, or large corporations. Thus, an ISP's goal is to deploy an IP-centric infrastructure capable of transporting the data exchanged between the connected customers and the Internet backbone.

### Application Requirements

As of publication of this book, the growing amount of Internet traffic requires the backbone trunks to be 2.5 Gbps or greater. The used routing platforms must scale up to the hundreds of gigabits of backplane capacity and must support 2.5 Gbps and 10 Gbps.

It must be possible to traffic engineer the data flows in the network to ensure efficient backbone resource utilization. A key point is also to be able to treat certain data with higher priority and to protect it just as quickly in failure scenarios.

As voice and multimedia applications are increasingly transported over IP, mechanisms must be available to guarantee certain levels of Quality of Service (QoS). These QoS levels incorporate definitions for characteristic transmission parameters, such as maximum transmission delay and jitter.

## Common Solutions

A typical ISP infrastructure is shown in Figure 5–14. The core network consists of several Core-POPs, typically located in major cities, interconnected with multiple 2.5-/10-Gbps Packet over SONET/SDH (POS) trunks. These trunks might utilize wavelengths provided by long-haul DWDM systems or directly by dark fiber, together with regenerators.

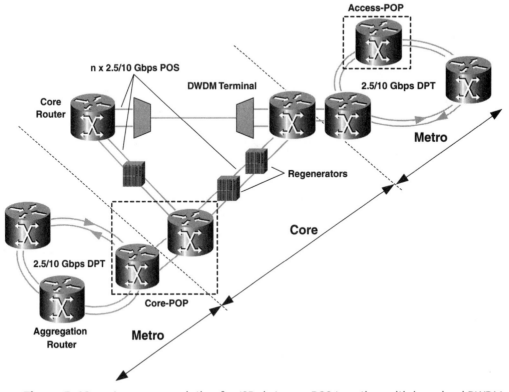

**Figure 5–14**    A common solution for ISPs is to use POS together with long-haul DWDM systems in the network core and DPT in the metro network

Long-haul DWDM systems can be typically considered in scenarios where the ISP is part of a telecommunications-affiliated group. The affiliate responsible for deploying the TDM infrastructure typically also deploys DWDM systems to boost the available capacity of the fiber plant. Thus, the ISP simply rents certain wavelengths to deploy the IP backbone.

In case of an alternative ISP focusing exclusively on IP services, there is no DWDM and TDM infrastructure implicitly required and, therefore, typically none deployed. A cost-effective solution in this kind of scenario is to rent dark fiber and use regenerators to be placed every 40–80 km to ensure proper long-distance transmission between the backbone routers.

In the metro network, DPT rings are used to aggregate the high amount of Access-POPs into the core network. The DPT rings might be built using dark fiber, which is typically widely available in the metro area of major cities or might be built on top of metro DWDM or high-capacity SONET/SDH networks. The use of ring topologies in the metro network reduces the amount of connections required to interconnect the Access- and Core-POPs.

Both POS and DPT technologies provide optimized IP transport over high-bandwidth interfaces. The decision whether to use POS or DPT cannot be answered in general. There are several aspects to be taken into account, such as already installed interfaces, network topology, and interface availability for the used router systems. Often, the decision depends on the historical evolution of an ISP network or on human preferences.

**Table 5–1**    Advantages and Disadvantages of POS and DPT Technologies for the Core and Metro Networks

| | POS | DPT |
|---|---|---|
| Bandwidth | (+) Easy upgrade possible by just adding another trunk between POPs | (+) SRP-fa algorithm guarantees fair share of bandwidth for each POP on the ring <br> (+) The spatial reuse functionality ensures efficient resource utilization <br> (-) A bandwidth upgrade requires all POPs on the ring to be upgraded |
| Traffic Engineering | (+)Aggregated data flows can be directly routed onto certain trunks | (-) DPT provides a shared media and, therefore, provides no implicit control of how the traffic is transmitted across the ring |
| Network Restoration | (+) MPLS-TE fast reroute makes restoration within less than 100 ms possible <br> (-) Without MPLS-TE fast reroute, restoration relies on routing protocol convergence (seconds) | (+) IPS provides protection within less than 50 ms and is, therefore, transparent to IP at layer 3 |

Table 5–1 should help to make the decision between the use of POS or DPT in the network. Basically, there are three areas of concern. The first is network bandwidth—how it is allocated and how it can be upgraded. DPT ensures that the available bandwidth on the ring is adequately shared between the POPs on the ring. Figure 5–15 shows an OC-48c/STM-16c DPT ring with five POPs and a diagram of the traffic flows inserted into the ring. As long as POP A is the only node sending traffic across the ring toward POP C, it can use the whole 2.5 Gbps. After a while, POP B also starts sending traffic across POP C, leading to congestion at the ring segment between POP B and POP C. The SRP fairness algorithm (SRP-fa) now throttles POP A down to 1.25 Gbps to ensure that

POP B also can insert 1.25 Gbps of traffic into the ring. In a third step, POP D starts inserting traffic toward POP B. Again, congestion appears at the segment between POP B and POP C, and, again, the SRP-fa throttles down the other POPs to ensure that POP D can insert its fair share (~ 800 Mbps) of bandwidth into the ring.

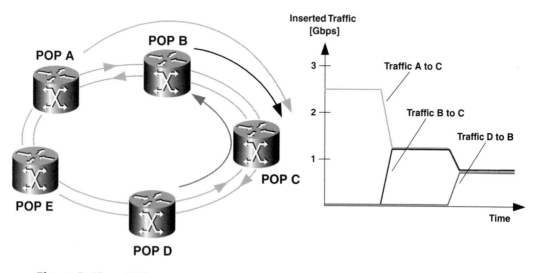

**Figure 5–15**    DPT ensures appropriate bandwidth sharing between the POPs on the ring

A drawback, on the other hand, is that an upgrade, for example, from 2.5 Gbps to 10 Gbps, must be done for the whole ring and, therefore, requires new interfaces for each POP on the ring. Figure 5–16 shows that POS has a significant advantage here. Each POS p-t-p link can be easily upgraded by exchanging, for instance, the 2.5-Gbps POS interface on both POPs with a 10-Gbps POS interface or by adding additional POS interfaces for creating parallel links and smoothly upgrading the bandwidth of the p-t-p link.

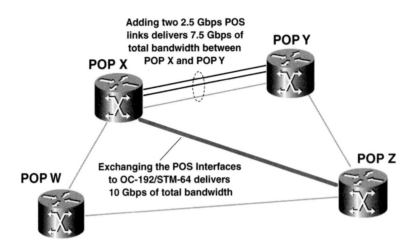

**Figure 5–16**    POS allows partial network bandwidth upgrades

Traffic engineering is the second area of concern. Using POS creates a mesh network topology with dedicated p-t-p links. MPLS Traffic Engineering (MPLS-TE) can be used to force traffic going across certain links. Using DPT means creating a shared media between the interconnected POPs. How the traffic flows on the ring is controlled by the SRP protocol mechanisms: topology discovery, SRP-fa, and Intelligent Protection Switching (IPS). The network operator cannot force the traffic going across the right or left side of the ring.

The third area where POS and DPT can significantly be differentiated is network restoration. The IPS functionality of DPT handles failures on the ring, such as fiber cuts or router failures, by wrapping the inner and outer rings around the failure and restores the network within the magic 50-ms boundary. In a POS network, no layer-2 protection is available; thus, the network has to be restored by the routing protocol. The restoration time depends on the size of the network and of the configured routing protocol timers, for example, the Hello packet interval and dead timer, etc., in case of the Open Shortest Path First (OSPF) protocol used for routing. To increase the level of network survivability, the fast-reroute functionality of MPLS-TE can be used. Backup tunnels can be preestablished around important links or nodes, which are used to reroute affected tunnels in failure scenarios. For details, refer to Chapter 3, "Optical Networking Technology Fundamentals."

Apart from the discussion of whether to use DPT or POS, MPLS-TE is commonly deployed in ISP networks to get more control of the connectionless, packet-based infrastructure. Considering the network architecture as shown in Figure 5–14, several tunnels between Core-POPs are typically set up. Each tunnel is carrying a certain aggregate of data flows. In the simplest scenario, that might be the entire traffic from the Core-POP toward the core network, which can be restored by the MPLS protection mechanisms using preestablished or on-demand backup tunnels.

Some ISPs apply more complex traffic engineering scenarios, as shown in Figure 5–17. They use tunnels to separate customer traffic in their core network and to deliver an appropriate service quality to their customers. For example, an ISP might have as a customer a large enterprise requiring a bidirectional, 500-Mbps "virtual IP pipe" from Access-POP A to Access-POP D. In addition, the customer wants this tunnel to be protected and requests a certain minimum delay across this pipe. To ensure this service level, the ISP is setting up one MPLS tunnel from Access-POPs A to D and one MPLS tunnel from Access-POPs D to A, each with a bandwidth of 500 Mbps. To ensure the required minimum delay, the ISP can explicitly define the best path along the tunnels are to be routed through the network. By adding these new tunnels to the list of protected tunnels for each backup tunnel in the network, protection is enabled and provided to the customer.

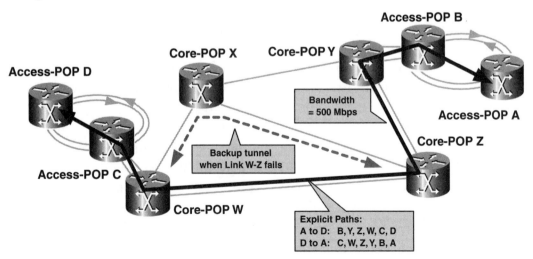

**Figure 5–17**    In more complex scenarios, MPLS-TE tunnels are used to ensure certain service levels to customers

In addition to MPLS-TE, QoS mechanisms are also commonly deployed in IP transport networks. The most common and most scalable approach for ISPs to deliver certain QoS levels is the *DiffServ architecture*.

Within the DiffServ architecture, traffic is classified at the edge of the network. This classification is either done at the customer's premise equipment (CPE) or at the Access-POP of the ISP. The process for assigning traffic to several Classes of Service (CoS) is also often referred as *coloring of packets*. Hence, as the traffic is then traveling through the network, each router can distinguish between important and nonimportant traffic and handle it appropriately by taking into account is CoS.

Figure 5–18 summarizes all processes required in a router to enable QoS in the IP network. After traffic has entered the router, it is policed and may be discarded or shaped. This allows the ISP to limit customers from sending above their defined traffic limit to their specified traffic volume. This traffic-conditioning mechanism contains several functions, such as metering, dropping, marking, shaping, and accounting. They are defined by standards and implemented in proprietary ways by different vendors. It should be understood that they are, in general, very CPU-intensive and also very complex to manage. This means that they should be deployed only in very strategic parts of a network, such as transition points from the service provider to enterprises or to other service providers. In general, it is not a good idea to conditioning them in the core of the network because the traffic should already be forwarded in the most suitable way.

In a second step, the defined traffic classes are separated into multiple queues. **Congestion avoidance** is applied to each queue separately. Weighted Random Early Discard (W-RED) is the industry standard for congestion avoidance and ensures that the depth of each queue stays within specified boundaries by randomly discarding packets in a proactive fashion.

In the third and last stage, the traffic is dequeued again and sent across the outgoing router interfaces, using certain **queuing mechanisms**, which are foremost described in standards and implemented by different vendors in different ways. Those mechanisms send high-priority traffic first and, therefore, ensure that delay and jitter is as specified in the CoS definition.

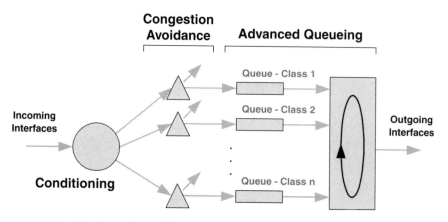

**Figure 5–18**    Each router applies congestion avoidance and scheduling mechanisms to ensure the defined QoS levels

## Case Example C: Next-Generation Carrier

Traditional service providers that started as telephony providers are faced with totally different challenges when designing their networks. The goal is to deploy a future-proof infrastructure with scalable capacity, delivering a broad set of services and a set of simplified provisioning processes that enable service velocity and operational cost reduction.

One of the biggest parts of the service portfolio is probably construed through a mix of "legacy" circuit-based low-bandwidth services, such as leased lines in the range of 32 Kbps up to 2 Mbps, and a set of high-bandwidth services, such as SONET/SDH circuits ranging from OC-3/STM-1 up to OC-192/STM-64.

The fastest growing part of the service portfolio is, of course, a set of data services based on IP over Ethernet or SONET/SDH and a set of IP or non-IP data services based on ATM.

Considering that enterprises commonly try to outsource their networks, a new set of storage services based on the mainframe technologies ESCON and FC provide a new and quickly emerging business opportunity for carriers when providing these services in their portfolios.

Of course, with the emergence of bit- and protocol-transparent DWDM systems and optical switching, technology-transparent optical services (often referred to as ***Clear Channel***) are to be seen more and more.

## Application Requirements

Having provided the long list of services that should be provided by the carrier's infrastructure, it is clear that next-generation carriers need a flexible solution based on several transmission technologies.

The applied transmission systems must provide several "low-order" TDM interface types, such as T1/E1 or DS3/E3, and several "high-order" TDM interface types, such as OC-n/STM-n. In addition 10-/100-/1000-Mbps Ethernet-type data interfaces and storage interface types, such as ESCON and FC, must be provided.

The solution must be very scalable and flexible to ensure a smooth migration path for services, e.g., from 100-Mbps Ethernet to Gigabit Ethernet. Particularly, in the metro part of the network, the potential to introduce DWDM technology must be provided.

In the core part of the network, efficient capacity utilization is the key. The bandwidth of each wavelength transmitted over the core DWDM systems should be exploited as far as possible, and dynamic connection provisioning and reoptimization mechanisms must ensure optimum lightpath placement. Dynamic mesh restoration is favored to minimize the amount of prereserved capacity for protection purposes.

To minimize operational costs, integrated network management for A–Z provisioning and uniform fault and performance monitoring is absolutely required in order to streamline and simplify network operation processes. In order to reduce ongoing costs, environmental requirements such as small footprint and low power consumption are also very important points of concern.

## Common Solutions

A way to address all these requirements is to use a mix of transmission systems, such as DWDM systems, wavelength routers, next-generation SONET/SDH systems, ATM switches, and IP routers (Figure 5–19). Each technology has its key functions and advantages when used in the right place in the network.

Very obvious is the use of long-haul DWDM systems in the core of the network. Commercially available long-haul DWDM systems typically provide up to 120 or 160 wavelengths in the 50–100-GHz ITU grid over SONET/SDH-based transponders at 2.5 Gbps, 10 Gbps, or 40 Gbps. The size of the optical spans between optical amplifier sites and the maximum transmission distance without electrical regeneration vary, depending on the fiber type, wavelength bit rates, as well as optical amplifying and dispersion compensation technologies used in the DWDM system.

To get the most efficient access to the bandwidth provided by DWDM, a mesh architecture is deployed in the core, and wavelength routers are used in the DWDM junctions. Wavelength routers typically provide up to 256 or more OC-48/STM-16 or OC-192/STM-64 interfaces. The wavelength routing protocol may be a proprietary solution or may be standards-based, but, in both cases, it ensures efficient bandwidth allocation through automated and dynamic A–Z provisioning. High network resilience in the core is furthermore achieved through the mesh restoration functionality.

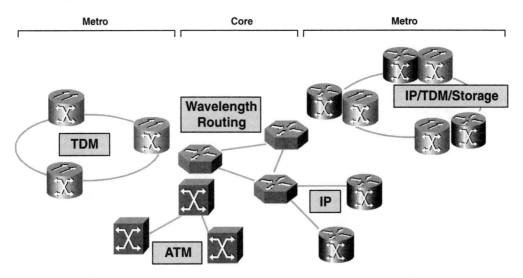

**Figure 5–19**   The next-generation network architecture integrates TDM and data services by using a mix of wavelength routing, IP, ATM, and SONET/SDH

In the metro part of the carrier's network, DWDM is also typically used to cope with the massive bandwidth demand in regional areas around major cities. Metro DWDM systems typically provide 32 wavelengths in the 200-GHz ITU grid over transparent transponders. This allows the transport of SONET/SDH, ATM, IP, and ESCON over the same type of DWDM transponder. Furthermore, once a metro DWDM system is installed, easy bandwidth upgrades are made possible in the metro by simply installing another transponder card.

As opposed to the trend in the core network, where optical switching and wavelength routing replaces traditional SONET/SDH ring infrastructures, SONET/SDH is still used for grooming circuit-based low-order traffic to the OC-192/STM-64 level in the metro network. This ensures, on the one hand, efficient core network resource utilization and, on the other hand, investment protection in the metro network because existing SONET/SDH multiplexers aggregating T1/E1, ADSL, and other low-bandwidth customer connections may be reused in the access network.

Another driver for using SONET/SDH in the metro is the existence of new, next-generation SONET/SDH systems. As opposed to the past, one system can now act as add/drop multiplexer (ADM), terminal multiplexer (TM), and digital cross-connect (DXC). This simplifies the network architecture, enables easy end-to-end "point-and-click" provisioning, therefore also streamlining network operation.

Even more compelling is the fact that Ethernet and ATM switching are integrated into these systems. Carriers can transparently carry Ethernet traffic across the metro network and give their customers the opportunity to use cheaper CPE; such as low-cost routers with Ethernet interfaces or even small Ethernet switches with integrated layer-3 routing functionality, as opposed to expensive middle or high-end routers with SONET/SDH-based interfaces.

To get a bit deeper understanding of the next-generation carrier architecture, the structure of typical Service POPs in the metro and core network is outlined now. Going from the edge of the carrier network toward the core, the Service Access-POP is the first step. Figure 5–20 shows a simplified architecture of a typical Service Access-POP aggregating the access links into the metro network.

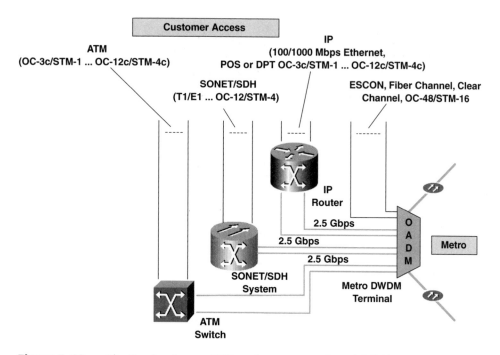

**Figure 5–20** The Service Access-POP performs grooming at 2.5 Gbps or even at the 10-Gbps level between the customer access and the metro network

The IP routers in the Service Access-POP provide a mix of access interface types, including POS and DPT interfaces—typically at speeds up to OC-12c/STM-4c—Ethernet and Fast Ethernet interfaces, as well as other channelized SONET/SDH interfaces for leased line aggregation ranging from T1/E1 up to OC-12/STM-4. The IP traffic is aggregated and commonly sent across POS or DPT interfaces at OC-48c/STM-16c speeds.

Next to the IP routers, ATM switches might also be part of the Service Access-POP. Running MPLS on the ATM switches enables the carrier to integrate the control planes for both the IP routers and the ATM switches. The ATM switches typically aggregate low-bandwidth data traffic, providing OC-3c/STM-1 and OC-12c/STM-4c access interfaces and OC-48c/STM-16c interfaces toward the metro DWDM system.

The third part complementing the "multiservice" capability of the Service Access-POP are SONET/SDH systems, which are aggregating the low-order TDM traffic coming from access interfaces, ranging from T1/E1 up to OC-

12/STM-4 onto OC-48/STM-16 interfaces connected to the metro DWDM systems.

As already mentioned, a metro DWDM system is used to connect the Service Access-POP to the metro network. Transparent transponders allow the seamless connection of IP routers, ATM switches, and SONET/SDH systems. In addition, high bit rate signals, such as IP customer connections using POS or DPT at OC-48c/STM-16c speeds, can be directly aggregated at the DWDM system if desired. Furthermore, storage services are provided through special ESCON or FC TDM transponder cards (also see "Case Example A: Next-Generation Storage Networks").

Going forward through the metro network, the next step is the Service Core-POP. Figure 5–21 shows a simplified architecture of a typical Service Core-POP connecting the metro network to the core network.

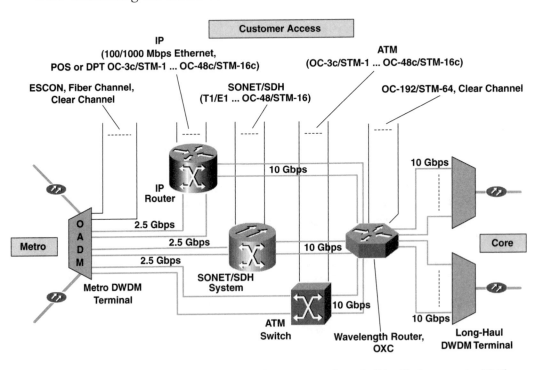

**Figure 5–21**   The Service Core-POP performs grooming of all traffic types onto 10-Gbps wavelengths between the core and metro networks

At the metro side of the Service Core-POP, a metro DWDM system terminates the multiple traffic streams coming from the Service Access-POPs and hands them over again to IP routers, ATM switches, or SONET/SDH systems. At the Service Core-POP, a second level of grooming is performed onto typically 10-Gbps signals for the multiple IP, ATM, or TDM signals, either coming across the metro network or being inserted directly at the Core-POP. On the core side of the Service Core-POP, the wavelength router is providing lightpaths typically carrying signals at 10 Gbps to the IP routers, ATM switches, or SONET/SDH systems. These lightpaths are routed across the attached long-haul DWDM systems through the core network to the desired destination Core-POP.

In the end, the result can be quite a complex network architecture, including several technologies, and might be quite difficult to manage in terms of service provisioning and network control. Because the carrier cannot deploy an IP-only optical transport infrastructure, multiple separated management systems for each technology have to be maintained. This fact is the driver for the control plane evolution seen in the industry. As discussed in previous chapters in greater detail, the MPLS-TE architecture is to be adopted and used for controlling not only IP and ATM infrastructures but also the connection setup, tear-downs, and protection functions in the optical transport infrastructure, including DWDM systems, wavelength routers or optical cross-connect (OXC) and SONET/SDH systems.

On the way to a unified control plane, there are some evolutionary milestones. In the first step, proprietary wavelength routing solutions are used in the meshed optical core network. With the standardized optical user-to-network interface (UNI), dynamic and automated interaction between the optical core and metro network elements is possible in the second step. IP routers and SONET/SDH systems can automatically request core network connection setups and tear-downs from wavelength routers, reducing the provisioning effort between the core and the metro network. The MPLS-based standard control plane solution shown in Figure 5–22 seamlessly integrates IP, ATM, and SONET/SDH network elements in the third step. Using MPLmS also opens the possibility to deploy the peer network model, which streamlines the network architecture to a single layer from a control perspective.

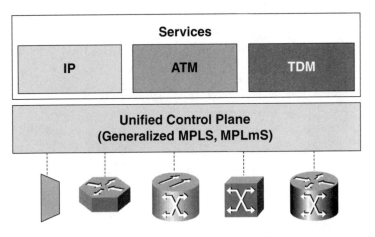

**Figure 5–22**    A unified optical control plane simplifies the network management and provisioning processes

## Summary

This chapter completed this work on the convergence of IP and optical technology with a practical view on optical network end-to-end design, together with three case examples describing the major applications in the field of OTNs.

We showed the evolution of traditional POPs to Service POPs as the service delivery point and the connection between users and services as well as the core and metro network. The core network is moving toward a mesh infrastructure and—in the metro network—we see a mix of various technologies, such as IP, ATM, and SONET/SDH, together with DWDM.

Case Example A showed the trend in the storage networking arena with a huge demand for metro DWDM and an integration of the traditionally separated storage networks with the IP data network.

Case Example B showed how ISPs adapt their network infrastructure to the quickly growing Internet traffic volume by using POS and DPT technology to tie the IP network directly onto the optical infrastructure.

Finally, Case Example C described the situation of a next-generation carrier whose goal is to streamline and simplify its complex network architecture, including IP, ATM, SONET/SDH, and DWDM infrastructures to deliver a broad set of services with minimized operational costs.

# Recommended Reading

[CSCO-10] Cisco Systems Inc. Whitepaper, *Cisco Introduces an IP-Based Optical Control Plane—IP-OCP*, 2001.

[IBM-1] IBM Redbook, *OS/390 MVS Parallel Sysplex Configuration*, Volume 1: Overview, January 1998.

[IETF-29] IETF-draft, draft-ietf-ips-iscsi-05.txt, *iSCSI*, work in progress, March 2001.

# Glossary

| | |
|---|---|
| 2-fiber BLSR | Two-Fiber Bidirectional Line-Switched Ring |
| AAL | ATM adaptation layer |
| Access-POP | access points of presence |
| ADM | add/drop multiplexer |
| AFI | authority and format identifier |
| AGC | automatic gain control |
| AIS | alarm indicator signal |
| APS | Automatic Protection Switching |
| AS | autonomous system |
| ASE | amplified spontaneous emissions |
| ATM | Asynchronous Transfer Mode |
| ATM_ARP | an ATM address resolution protocol |
| B2B | business to business |
| B2C | business to consumer |
| BER | bit error rate |
| BFP | big fat pipes |
| BGP | Border Gateway Protocol |
| B-ICI | B-ISDN intercarrier interface |
| BLSR | Bidirectional Line-Switched Ring |
| CAPEX | capital expense |
| CAR | committed access rate |
| CD | chromatic dispersion |
| CE | customer edge |
| CLIP | Classical IP |

| | |
|---|---|
| CLP | cell loss priority |
| CoS | Class of Service |
| CPE | customer's premise equipment |
| CPM | cross-phase modulation |
| CPU | central processing unit |
| CRC | cyclic redundancy check |
| CS | convergence sublayer |
| CSPF | Constrained Shortest Path First |
| DCC | data country code |
| DCN | digital communication network |
| DEMUX | demultiplexer |
| DPT | Dynamic Packet Transport |
| DS | dispersion-shifted |
| DSF | dispersion-shifted fiber |
| DSL | Digital Subscriber Line |
| DSP | domain-specific part |
| DWDM | Dense Wavelength Division Multiplexing |
| DXC | digital cross-connect |
| EDFA | erbium-doped fiber amplifier |
| EDFFA | erbium-doped fluoride fiber amplifier |
| Edge-LSR | edge label switch router |
| ERO | `EXPLICIT_ROUTE` object |
| ESCON | Enterprise Systems Connection |
| FA | forwarding adjacency |
| FC | Fiber Channel |
| CF | Coupling Facility |
| FCS | frame check sequence |
| FDDI | Fiber Distributed Date Interface |
| FEC | forwarding equivalent class |
| FR | Frame Relay |
| FS | forced switch |
| FWM | four-wave mixing |
| GDPS | Geographic Dispersed Parallel Sysplex |
| HDLC | Highspeed Data Link Control Protocol |
| I | idle |

| | |
|---|---|
| ICD | international code designator |
| IDI | initial domain identifier |
| IEEE | Institute of Electrical and Electronics Engineering |
| IETF | Internet Engineering Task Force |
| IGP | Interior Gateway Protocol |
| IP | Internet Protocol |
| IPS | Intelligent Protection Switching |
| iSCSI | SCSI transport over IP |
| IS-IS | Intermediate System to Intermediate System protocol |
| ISO | International Organization for Standardization |
| ISP | Internet service provider |
| ISUP | ISDN user part |
| ITU | International Telecommunication Union |
| LAN | Local Area Network |
| LANE | LAN emulation |
| LCP | Link Control Protocol |
| LDP | Label Distribution Protocol |
| LFIB | label forwarding information base |
| LmFIB | lambda forwarding information base |
| LI | length indication |
| LIB | label information base |
| LmSC | lambda signaling controller |
| LOH | line overhead |
| LOS | loss of signal |
| LSA | link state advertisement |
| LSP tunnel | label-switched path tunnels |
| LSP | label-switched path |
| LSR | label switch router |
| LTE | line terminal equipment |
| MAC | media access control |
| MAM | maximum allocation multiplier |
| MAN | metropolitan area network |
| MD | modal dispersion |
| MEMS | Microelectro mechanical systems |
| MIB | Management Information Base |

| | |
|---|---|
| MII | media-independent interface |
| MMF | multimode fiber |
| MP-BGP | Multiprotocol BGP |
| MPLmS | Multiprotocol Lambda Switching |
| MPLS | Multiprotocol Label Switching |
| MPLS-TE | MPLS Traffic Engineering |
| MS | manual switch |
| MSOH | multiplex section overhead |
| MSP | Multiplex Section Protection |
| MS-SPRing | Multiplex Section Shared Protection Ring |
| MTU | maximum transmission unit |
| MUX | multiplexer |
| NBMA | nonbroadcast multiaccess |
| NCP | Network Control Protocol |
| NFOEC | National Fiber Optic Engineers Conference |
| NHLFE | next-hop label forwarding entry |
| NHRP | Next Hop Resolution Protocol |
| NNI | network node interface |
| NOC | Network Operation Center |
| NRZ | nonreturn to zero |
| NSAP | network service access point |
| NZD | nonzero dispersion fibres |
| OADM | optical add/drop multiplexer |
| OAM&P | operation, administration, maintenance, and provisioning |
| O-APS | Optical Automatic Protection Switching |
| O-BLSR | Optical Bidirectional Line Switched Rings |
| O-BLSR | Wavelength Bidirectional Line-Switched Rings |
| OC-1 | optical carrier 1 |
| OCH trail | optical channel trail |
| OCH | optical channel |
| OIF | Optical Internetworking Forum |
| OLA | optical line amplifier |
| O-LSP | optical LSP |
| OMS trail | optical multiplex section trail |
| OMS | optical multiplex section |

| | |
|---|---|
| O-MSP | Optical Multiplex Section Protection |
| OMS-SPRing | Optical Multiplex Section Shared Protection Ring |
| ON | optical networking |
| ONE | optical network element |
| O-NNI | optical network-to-network interface |
| OPEX | operational expenses |
| OSC | optical supervisory channel |
| O-SNCP | Optical Subnetwork Connection Protection |
| OSI | Open Systems Interconnection |
| O-LSP | optical label-switched path |
| OSNR | optical signal to noise ratio |
| OSPF | Open Shortest Path First protocol |
| OSU | optical switch unit |
| OTM | optical transport module |
| OTN | optical transport network |
| OTS trail | optical transmission section trail |
| OTS | optical transmission section |
| OTU | optical transmission unit |
| OTUG-n | OTU group of the order n |
| O-UNI | optical user-to-network interface |
| O-UPSR | Optical Unidirectional Path Switched Rings |
| O-UPSR | Wavelength Unidirectional Path Switching Ring |
| OXC | optical cross-connect |
| P | provider |
| P/F | poll/final |
| PDFA | praseodynamium-doped fiber amplifier |
| PDH | Plesiochronous Digital Hierarchy |
| PDU | protocol data unit |
| PE | provider edge |
| PHY | physical |
| PMD | polarization-mode dispersion |
| PNNI | private network-to-network interface protocol |
| POH | path overhead |
| POP | point of presence |
| POS | Packet over SONET/SDH |

| | |
|---|---|
| PPP | Point-to-Point Protocol |
| PT | payload type |
| p-t-p | point-to-point |
| PVC | permanent virtual connection |
| QoS | Quality of Service |
| ROI | return of investment |
| RPR | Resilient Packet Ring |
| RSOH | regenerator section overhead |
| RSVP | Resource Reservation Protocol |
| RSVP/LDP | Resource Reservation Protocol / Label Distribution Protocol |
| RSVP-TE | RSVP with Traffic Engineering Extensions |
| SAR | segmentation and reassembly |
| SBS | stimulated Brillouin scattering |
| SCSI | Small Computer System Interface |
| SNCP | Subnetwork Connection Protection |
| SD | signal degrade |
| SDH | Synchronous Digital Hierarchy |
| SDL | Simple Data Link |
| SDU | service data unit |
| SEAL | simple and efficient adaptation layer |
| SF | signal failure |
| SLA | Service Level Agreement |
| SMDS | Switched Multimegabit Data Service |
| SMF | single-mode fiber |
| SN | sequence number |
| SNP | sequence number protection |
| SNR | signal-to-noise ratio |
| SOH | section overhead |
| SONET | Synchronous Optical Network |
| SPF | Shortest Path First |
| SPM | self-phase modulation |
| SRP | Spatial Reuse Protocol |
| SRP-fa | SRP fairness algorithm |
| SRS | stimulated Raman scattering |
| STA | Spanning Tree algorithm |

| | |
|---|---|
| STM-1 | synchronous transport module 1 |
| STS-1 | synchronous transport signal 1 |
| SVC | switched virtual connection |
| TDM | Time Division Multiplexing |
| TLV | Type/Length/Value |
| TM | terminal multiplexer |
| TOS | Type of Service |
| TTL | time to live |
| TUG | transport unit group |
| UI | unnumbered information |
| UNI | user-to-network interface |
| UNI-C | UNI client-side signaling functionality |
| UNI-N | UNI network-side signaling functionality |
| UPSR | Unidirectional Path Switched Ring |
| VBR | variable bit rate |
| VC | virtual circuit |
| VC | virtual container |
| VCI | virtual channel identifier |
| VoIP | Voice over IP |
| VP | virtual path |
| VPI | virtual path identifier |
| VPN | Virtual Private Network |
| VT | virtual tributary |
| VTG | VT group |
| WADM | wavelength add/drop multiplexer |
| WAN | Wide Area Network |
| WDM | Wavelength Division Multiplexing |
| WLP | wavelength path |
| WR | wavelength router |
| WRC | wavelength routing controller |
| W-RED | Weighted Random Early Discard |
| WT | wavelength terminal |
| WTR | wait to restore |
| WXC | wavelength cross-connect |
| XOR | exclusive-or |

# Notes

[AF-1] ATM Forum Specification, af-lane-0021.000, *LAN Emulation over ATM Version 1.0*, January 1995.

[AF-2] ATM Forum Specification, af-uni-0010.002, *ATM User-Network Interface Specification V3.1*, 1994.

[AF-3] ATM Forum Specification, af-sig-0061.000, *UNI Signaling 4.0*, July 1996.

[ALC-1] Alcatel Technology Paper, *Today's Optical Amplifiers—The Cornerstone of Tomorrow's Optical Layer*, Thomas Fuerst.

[BELL-1] Bellcore GR-253-CORE, *Synchronous Optical Network (SONET) Transport Systems: Common Generic Criteria*, Issue 2, December 1995 (Revision 1, December 1997).

[CIENA-1] CIENA Corp. Whitepaper, *Fundamentals of DWDM*.

[CIENA-2] CIENA Corp. Whitepaper, *The Evolution of DWDM*.

[CIENA-3] CIENA Corp. Whitepaper, *The New Economics of Optical Core Networks*, Sept. 1999.

[CNET-1] CANARIE Inc., *Architectural and Engineering Issues for Building an Optical Internet*, draft, Bill St. Arnaud, September 1998.

[CSCO-1] Cisco Systems Inc. Whitepaper, *Scaling Optical Data Networks with Wavelength Routing*, 1999.

[CSCO-2] Cisco Systems Inc. Whitepaper, *Cisco's Packet over SONET/SDH (POS) Technology Support*, February 1998.

[CSCO-3] Cisco Systems Inc. Whitepaper, *Dynamic Packet Transport Technology and Applications Overview*, January 1999.

[CSCO-4] Networkers Conference—Session 606, *Advanced Optical Technology Concepts*, Vienna, October 1999.

[CSCO-5] Cisco Systems Inc. Whitepaper, *Cisco Optical Internetworking*, 1999.

[CSCO-6] Cisco Systems Inc. Whitepaper, *Designing ATM MPLS Networks*, Jeremy Lawrence, April 1999.

[CSCO-7] Cisco Systems Inc. Whitepaper, *ATM Internetworking*, March 1995.

[CSCO-8] Cisco Systems Inc. Whitepaper, *Building High-Speed Exchange Points*, Revision 1.2, Paul Ferguson, March 1997.

[CSCO-9] Cisco Systems Inc., *Cisco IOS 12.0 QoS Configuration Guide*.

[CSCO-10] Cisco Systems Inc. Whitepaper, *Cisco Introduces an IP-Based Optical Control Plane—IP-OCP*, 2001.

[IBM-1]   IBM Redbook, *OS/390 MVS Parallel Sysplex Configuration*, Volume 1: Overview, January 1998.

[IEEE-1] IEEE Communications Magazine, *Future Photonic Transport Networks Based on WDM Technologies*, Hiroshi Yoshimura, NTT Optical Network Systems Laboratories, February 1999.

[IETF-1] RFC 2615, *PPP over SONET/SDH*, June 1999.

[IETF-2] RFC 1661, *The Point-to-Point Protocol (PPP)*, July 1994.

[IETF-3] RFC 1662, *PPP in HDLC-Like framing*, July 1994.

[IETF-4] IETF-draft, draft-ietf-pppext-pppoversonet-update-00.txt, *PPP over SONET/SDH*, work in progress, February 1999.

[IETF-5] IETF-draft, draft-merchant-pppext-sonet-sdh-00.txt, *PPP over SONET (SDH) at Rates from STS-1 (AU-3) to STS-192c (AU-4-64c/STM-64)*, work in progress, November 1998.

[IETF-6] IETF-draft, draft-ietf-mpls-arch-02.txt, *Multiprotocol Label Switching Architecture*, work in progress, July 1998.

[IETF-7] IETF-draft, draft-ietf-mpls-traffic-eng-00.txt, *Requirements for Traffic Engineering over MPLS*, work in progress, October 1998.

[IETF-8] RFC 2892, *The SRP MAC Layer Protocol, Status Informational*, August 2000.

[IETF-9] RFC2558, *Definitions of Managed Objects for the SONET/SDH Interface Type*, March 1999.

[IETF-10] IETF-draft, draft-jedrysiak-srp-mib-00.txt, *SRP MIB*, work in progress.

[IETF-11] IETF-draft, draft-ip-optical-framework-00.txt, *IP over Optical Networks—A Framework*, work in progress, March 2000.

[IETF-12] IETF-draft, draft-chaudhuri-ip-olxc-control-00.txt, *Control of Lightpaths in an Optical Network*, work in progress, February 2000.

[IETF-13] IETF-draft, draft-awduche-mpls-te-optical-01.txt, *Multi-Protocol Lambda Switching: Combining MPLS Traffic Engineering Control with Optical Crossconnects*, work in progress, November 1999.

[IETF-14] IETF-draft, draft-basak-mpls-oxc-issues-01.txt, *Multi-Protocol Lambda Switching: Issues in Combining MPLS Traffic Engineering Control with Optical Crossconnects*, work in progress, November 1999.

[IETF-15] IETF-draft, draft-kompella-ospf-ompls-extensions-00.txt, *OSPF Extensions in Support of MPL(ambda)S*, work in progress, July 2000.

[IETF-16] IETF-draft, draft-kompella-isis-ompls-extensions-00.txt, *IS-IS Extensions in Support of MPL(ambda)S*, work in progress, July 2000.

[IETF-17] IETF-draft, draft-katz-yeung-ospf-traffic-01.txt, *Traffic Engineering Extensions to OSPF*, work in progress, October 1999.

[IETF-18] IETF-draft, draft-kompella-mpls-optical-00.txt, *Extensions to IS-IS/ OSPF and RSVP in Support of MPL(ambda)S*, work in progress, February 2000.

[IETF-19] IETF-draft, draft-rstb-optical-signaling-framework-00.txt, *Signaling Framework for Automated Provisioning and Restoration of Paths in Optical Mesh Networks*, work in progress, March 2000.

[IETF-20] IETF-draft, draft-fan-mpls-lambda-signaling-00.txt, *Extensions to CR-LDP and RSVP-TE for Optical Path Set-Up*, work in progress, March 2000.

[IETF-21] IETF-draft, draft-ietf-mpls-generalized-signaling-00.txt, *Generalized MPLS—Signaling Functional Description*, work in progress, October 2000.

[IETF-22] IETF-draft, draft-ietf-mpls-lsp-hierarchy-00.txt, *LSP Hierarchy with MPLS TE*, work in progress, July 2000.

[IETF-23] IETF-draft, draft-ietf-isis-traffic-01.txt, *IS-IS Extensions for Traffic Engineering*, work in progress, May 1999.

[IETF-24] IETF-draft, draft-ietf-mpls-rsvp-lsp-tunnel-05.txt, *RSVP-TE: Extensions to RSVP for LSP Tunnels*, work in progress, February 2000.

[IETF-25] IETF-draft, draft-ietf-mpls-cr-ldp-03.txt, *Constraint-Based LSP Setup Using LDP*, work in progress, September 1999.

[IETF-26] RFC2370, *The OSPF Opaque LSA Option*, July 1998.

[IETF-27] IETF-draft, draft-hsmit-mpls-igp-spf-00.txt, *Calculating IGP Routes over Traffic Engineering Tunnels*, work in progress, December 1999.

[IETF-28] IETF-draft, draft-makam-mpls-protection-00.txt, *Protection/Restoration of MPLS Networks*, work in progress, October 1999.

[IETF-29] IETF-draft, draft-ietf-ips-iscsi-05.txt, *iSCSI*, work in progress, March 2001.

[IETF-30] RFC1577, *Classical IP and ARP over ATM*, January 1994.

[ITU-1] ITU-T Draft Recommendation G.871, *Framework for Optical Networking Recommendations*, October 1998.

[ITU-2] ITU-T Draft Recommendation G.872, *Architecture of Optical Transport Networks*, July 1998.

[ITU-3] ITU-T Recommendation G.707, *Network Node Interface for the Synchronous Digital Hierarchy (SDH)*, March 1996.

[ITU-4] ITU-T Recommendation G.692, *Optical Interfaces for Multichannel Systems with Optical Amplifiers*, March 1996.

[ITU-5] ITU-T Recommendation M.495, *Maintenance: International Transmission Systems*, 1993.

[ITU-6] ITU-T Recommendation G.652, *Characteristics of a Single-Mode Optical Fiber Cable*, 1993.

[ITU-7] ITU-T Recommendation G.653, *Characteristics of a Dispersion-Shifted Single-Mode Optical Fibre Cable*, 1993.

[JUNP-1] Juniper Networks Whitepaper, *Traffic Engineering for the New Public Network*, January 1999.

[JUNP-2] Juniper Networks Whitepaper, *RSVP Signalling Extensions for MPLS Traffic Engineering*, August 1999.

[LUCT-1] Lucent Technologies Whitepaper, *Multi-Layer Survivability*, J. Meijen, E. Varma, R. Wu, Y. Wang, 1999.

[MINO-1] McGraw-Hill Companies, *IP Applications with ATM*, J.Amoss, Ph.D., D. Minoli, May 1998.

[MUKH-1] McGraw-Hill, *Optical Communication Networks*, Biswanath Mukherjee, 1997.

[NFOEC-1] Technical Paper, *"IP over WDM" the Missing Link*, P. Bonenfant, A. Rodrigues-Moral, J. Manchester, Lucent Technologies, A. McGuire, BT Laboratories, September 1999.

[NFOEC-2] Technical Paper, *Packet over SONET and DWDM—Survey and Comparison*, D. O'Connor, Li Mo, E. Catovic, Fujitsu Network Communications Inc., September 1999.

[NFOEC-3] Technical Paper, *Position, Functions, Features and Enabling Technologies of Optical Cross-Connects in the Photonic Layer*, P. A. Perrier, Alcatel, September 1999.

[NORT-1] Nortel Whitepaper, *SONET 101*.

[OIF-1] OIF, *The Optical Internetworking Forum Kick-Off Meeting Presentation*, May 1998.

[OIF-2] OIF Contribution, *A Proposal to Use POS as Physical Layer up to OC-192c*, *OIF99.002.2*, January 1999.

[OIF-3] OIF Contribution, *IP Centric Control and Signaling for Optical Lightpaths*, January 2000.

[OIF-4] OIF Contribution, *User Network Interface (UNI) 1.0 Signaling Specification*, *OIF2000.125.3*, December 2000.

[RAM-1] Morgan Kaufmann Publishers Inc., *Optical Networks—A Practical Perspective*, Rajiv Ramasawi, Kumar N. Sivarajan, 1998.

[TELL-1] Tellium Whitepaper, *WDM Optical Network Architectures for a Data-Centric Environment*, Krishna Bala.

# Index

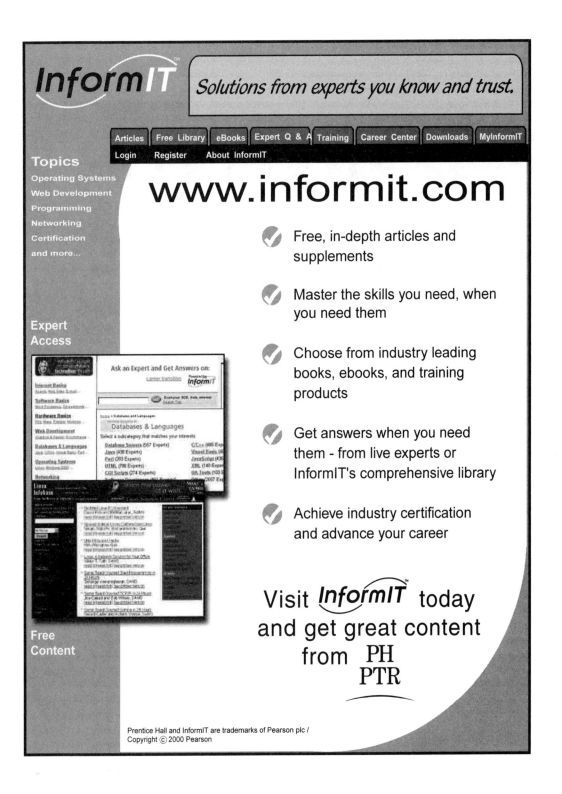